Advanced Electrochemical pH Sensing Technologies

Scientific fundamentals and applications

Online at: https://doi.org/10.1088/978-0-7503-6079-1

Advanced Electrochemical pH Sensing Technologies

Scientific fundamentals and applications

Libu Manjakkal

*School of Computing and Engineering and the Built Environment,
Edinburgh Napier University, Merchiston Campus, EH10 5DT, Edinburgh, UK*

IOP Publishing, Bristol, UK

ISBN   978-0-7503-6079-1 (ebook)
ISBN   978-0-7503-6077-7 (print)
ISBN   978-0-7503-6080-7 (myPrint)
ISBN   978-0-7503-6078-4 (mobi)

DOI   10.1088/978-0-7503-6079-1

Version: 20251101

IOP ebooks

British Library Cataloguing-in-Publication Data: A catalogue record for this book is available from the British Library.

Published by IOP Publishing, wholly owned by The Institute of Physics, London

IOP Publishing, No.2 The Distillery, Glassfields, Avon Street, Bristol, BS2 0GR, UK

US Office: IOP Publishing, Inc., 190 North Independence Mall West, Suite 601, Philadelphia, PA 19106, USA

This work is dedicated to my wife, Gopika Libu and my kids, Gokul Manjakkal and Dheerav Manjakkal, for providing me extraordinary support during the writing of this book. They sacrificed their personal time when I was busy with writing this book. They are very understanding about my career importance and they make my life more meaningful and happy.

Contents

Foreword

I am delighted to write this foreword for Dr Libu Manjakkal, Associate Professor, Edinburgh Napier University, for his excellent book *Advanced Electrochemical pH Sensing Technologies*. I have known Libu for more than five years during his research work at the University of Glasgow, where I was his mentor and line manager. He completed his PhD thesis in the field of electrochemical metal oxide-based pH sensors in the framework of the SENSEIVER project no 289481 of FP7 Marie Skłodowska-Curie Program (ITN) at the Institute of Electron Technology (ITE), Poland. Libu was working in the field of electrochemical sensors and supercapacitors for wearable applications. He has an excellent knowledge of electrochemical sensors for various applications, with an excellent track record in project initiation and delivery, Training and supervision, and in converting outputs into high-level publications.

As we know, the measurement of pH value has a significant influence on our daily life, which provides essential information related to many chemical reactions. The technological advances in the past few decades have led to the development of new pH sensors. So far, there are different methods exploited for measuring pH of a solution, both electrochemical and non-electrochemical. In his book, Libu provides a fundamental discussion on various approaches to electrochemical pH sensing technologies including development and their applications. This book covers the importance of materials, charcaterisation, various fabrication approaches, and the need for innovations in this area.

Considering the deep knowledge of Libu in electrochemical pH sensors, he is the ideal person to write this book, and I personally believe this book will provide readers an excellent grounding in the fundmentals, and state-of-the-art in this area. Students who plan to carry out research studies in this area will find this book to be very useful.

I wish all the best for Libu Manjakkal in the publication of this book.

Professor Richard Hogg
College of Engineering and Physical Sciences
Aston University

Foreword

It is with great pleasure that I write this foreword to the book *Advanced Electrochemical pH Sensing Technologies*. This book offers a comprehensive discussion of materials-based electrode designs for pH sensors, evaluating their applications across various fields and highlighting their suitability for emerging and future technologies. It also offers a detailed overview of rigid, flexible, and stretchable pH sensors, examining their capabilities in various environments.

This book presents a comprehensive selection of state-of-the-art materials, designs, and types of pH sensors, highlighting their significant potential in environmental monitoring, agriculture, food quality assessment, wearable devices, and aquaculture applications. It serves as a valuable resource for understanding both the fundamental and applied aspects of pH sensing technology.

The book discusses major analytical tools and electrochemical techniques—such as cyclic voltammetry, potentiometry, and electrochemical impedance spectroscopy —that are essential for investigating pH sensing mechanisms and performance.

A wide range of materials and fabrication approaches is explored, covering innovative sensor architectures and their impact on sensitivity, selectivity, and stability. Each chapter provides an in-depth analysis of various sensing materials, their interactions with interfering ions or analytes, and the factors affecting their response accuracy. Additionally, the book addresses key considerations such as material cost, environmental impact, and toxicity, offering a balanced perspective on performance versus sustainability.

Dr Manjakkal has a good track record in material science and electrochemical technologies, especially pH sensors. His expertise in this area strongly supports his ability to complete this book to an excellent quality. By combining theoretical insights with practical design strategies, this book aims to guide PhD researchers, postdoctoral researchers, process engineers, and master and undergraduate students working in the areas of electrochemistry, nanotechnology, wearable technology, water quality monitoring, chemical engineering, aquaculture/agriculture, and materials science.

Professor Suresh C Pillai
Nanotechnology and Bio-Engineering Division
Atlantic Technological University (ATU), Ireland

Preface

Technological advances over the past few decades have led to the development of new pH sensors. Currently, various methods are employed to measure the pH of a solution, including both electrochemical and non-electrochemical techniques. Among these, the electrochemical method is the most commonly available and reliable for measuring the pH value of a solution. Electrochemical pH sensors are in high demand across many applications such as food processing, health monitoring, agriculture, nuclear sectors, and water quality monitoring. Several factors, including fabrication methods and material properties, influence the sensing performance of electrochemical pH sensors. A fundamental understanding of material properties, electrode reactions, and measurement techniques will facilitate the development of new pH sensors.

I conducted my PhD studies in electronic engineering through the prestigious Marie Skłodowska-Curie ITN programme, specialising in electrochemical pH sensors development at the Institute of Electron Technology, Poland [New name of the institute—Łukasiewicz Research Network—Institute of Microelectronics and Photonics, Poland]. My research focuses on materials and electrode fabrication, and their implementation in electrochemical sensors and energy storage devices for healthcare and pollution monitoring. In the field of sensors, I have developed multiple sensors for monitoring pH, glucose, ammonia, and other ions. The pH sensors designed for water quality monitoring and healthcare applications have attracted significant interest. My expertise in this area gives me the confidence to author my first book on pH sensing technologies.

Acknowledgments

First and foremost, I would like to express my sincere gratitude to my PhD supervisor, Professor Dorota Szwagierczak, for her boundless support and encouragement for my research career and personal development. Her broad knowledge of the subject and in technical details have been truly supportive and motivational for me to finish my PhD studies in pH sensors. The lessons I got from my Professor are a valuable resource for my life. I have also greatly appreciated my auxiliary supervisor, Dr Jan Kulawik, for his technical support during my PhD studies.

I have published many pH sensors research and review articles, and I sincerely appreciate all my coauthors for their excellent support. I learned the technology of pH sensing from the Kraków Division of the Institute of Electron Technology, Poland, and my special thanks go to all my colleagues from there. I would like to express my sincere gratitude to my colleagues at Edinburgh Napier University, especially my students and team members. Also, I wish to express my gratitude to the colleagues from my previous institute University of Glasgow, UK, NOVA School of Science and Technology in Lisbon, Portugal, and C-MET, Kerala, India for their valuable support. I am especially thankful to my mentors: Professor Dorota Szwagierczak, Professor Richard Hogg, Professor Suresh C Pillai, Professor Gin Jose, Professor Luis Pereira, Dr Raghu Natarajan and Dr S N Potty. I gratefully acknowledge the European Commission in the framework of the FP 7 project SENSEIVER, grant number 289481, for financially supporting my PhD work. As a Marie Curie Early Stage researcher, I had the chance to collaborate under secondment programs with other universities and institutes, such as the University of Novi Sad, the Faculty of Technical Sciences in Serbia, Vienna Technical University in Austria and NORTH Point Ltd, Subotica, in Serbia.

Finally, and most importantly, I would like to thank my family and friends. I am a very lucky person to have such a wonderful family.

Author biography

Libu Manjakkal

Dr Libu Manjakkal (PhD, FHEA, MRSC) is an Associate Professor at Edinburgh Napier University, UK and works as group leader for the Sustainable Materials Research and Technologies (SMART) Group@Napier.

I was born the elder son of Babu Manjakkal and Leela Chenoottil, in a small village called Peringave, Malappuram, Kerala, India. I completed my preliminary and high school near Peringave. I received BSc (2006), MSc (2008) degrees in physics from Calicut University and Mahatma Gandhi University, India. From 2009 to 2012, I was with CMET, Thrissur, India, as project staff and in 2012, I worked as a researcher at NOVA School of Science and Technology in Lisbon, Portugal. I completed a PhD in electronic engineering from the Institute of Electron Technology, Poland (2012–15) (Marie Curie ITN Program). After completion of the PhD, I worked for one year as a research engineer in the same institute. Between 2016 and 2022, I was a post-doctoral fellow at the University of Glasgow in various roles, including research assistant, research associate and scientific project manager of a Marie Curie ITN project. Currently, my research focuses on material synthesis, wearable energy storage, electrochemical sensors, supercapacitors, electrochromic energy storage and energy-autonomous sensing systems development. I authored/co-authored more than 75 peer-reviewed papers (55+ journals). I was shortlisted for important awards such as the European Commission for Marie Curie Awards in the 'Promising Research Talent' category, the 'Outstanding Young Scientists Scholarship Award' from the Polish Ministry of Higher Education and the RIENG 2024 Young Investigator Award. I am currently also working as the editor of *Chemical Engineering Journal* and *Results in Engineering Journal*, Elsevier.

Chapter 1

Overview of pH sensing technologies

1.1 Introduction

The measurement of pH, which enables the determination of a solution's acidity or basicity, is significant in many chemical and biological reactions, as well as in environmental sciences. The pH value measured in the range of 0–14 has a significant influence on our daily life. For example, in food processing, it controls protein denaturation [1, 2], colour of food [3], gelification [4, 5], growth and mortality of microorganisms [6, 7], and the germination or inactivation of bacterial spores [8, 9]. The pH level also plays a critical role in the diagnosis of certain diseases, and it influences the shape of proteins in our blood [10–12]. The pH measurements are important in medicine, biology, chemistry, agriculture, forestry, food science, environmental science, chemical engineering and in many other applications. It is one of the major factors in the field of environmental pollution monitoring and industrial wastewater measurement [13–15]. The rate of chemical reactions and the solubility of chemicals or biomolecules are dependent on pH value. To optimize the desired reactions or to prevent unwanted reactions, the pH value must be controlled. Hence, the measurement of pH plays a superior role in a balanced ecosystem. The monitoring of pH determines valuable information related to the chemical composition of the environment and health. Technological advances in the past few decades have led to the development of new pH sensors, especially for continuous monitoring. To date, various methods have been employed for measuring the pH of a solution, including both electrochemical and non-electrochemical approaches [16–21], as illustrated in figure 1.1.

The electrochemical method is the most commonly used and reliable technique for measuring the pH of a solution [16, 17, 22]. Based on the electrochemical concept, there are different techniques employed for the fabrication of pH sensors, including the use of potentiometric, chemi-resistive, impedancemetric/conductimetric and ion-sensitive field effect transistors (ISFETs)-based sensors [16, 17, 23–26]. Many pH sensors are reported or developed individually or integrated with other sensors, including pH and glucose

doi:10.1088/978-0-7503-6079-1ch1

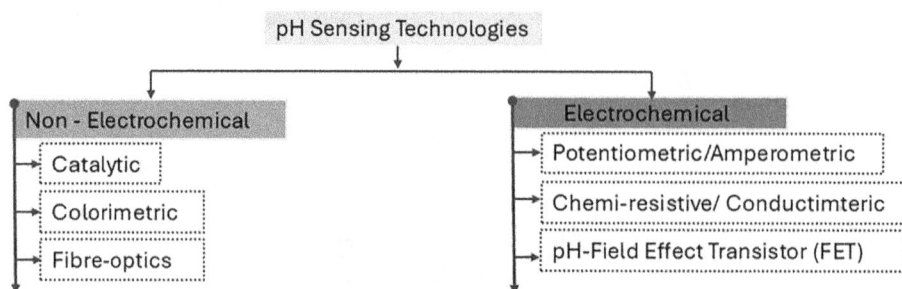

Figure 1.1. Non-electrochemical and electrochemical pH sensing technologies.

sensors [27], pH and temperature sensor [28], pH and Na monitoring [29], pH and urea monitoring [30], etc. These sensors predict the sensing mechanism and reveal the independence of each parameter's variation in health monitoring, environmental pollution determination or agriculture and food technologies. In addition to electro-chemical sensors, magnetoelastic sensors [31, 32] and micromechanical cantilever-based sensors [33, 34] are also implemented for measuring the pH of a solution. These are developed based on the sensing performances of the sensitive electrodes, which are prepared by various methods, miniaturised design of the sensors and the working mechanism. A few of the recently developed pH sensors are summarised in figure 1.2, which shows the different classes of sensors implemented for various applications [15, 35–39]. The sensitivity, selectivity, response time, stability and cost of fabrication are very important in the design of pH sensors.

The electrochemical pH sensor based on the glass electrode has been traditionally used for measuring pH due to its accuracy, excellent stability over a wide pH range, good response close to Nernstian behavior, and long lifetime [18, 40, 41]. However, the glass electrode exhibits several disadvantages, including mechanical fragility, difficulty in miniaturisation, chemical instability in corrosive systems, and a distinct dependence on pressure and temperature [42]. Furthermore, the glass electrode has some limitations for application in food processing and online monitoring of solution pH due to a lack of robustness. Researchers have been attempting to overcome these challenges for a long time and develop new pH measurement systems (a few of which are summarised in figure 1.2, which could also facilitate the development of a new generation of wireless pH sensors for online monitoring of solutions. Hence, there is a strong demand for the development and investigation of new low-cost sensitive materials, new sensor designs and measurement methods for next-generation mini-aturised pH sensors. Moreover, further efforts are necessary to elucidate the pH sensing mechanism. Considering the importance, this chapter discusses the definition of pH, the important range in different applications and sensing mechanisms, which are summarised for electrochemical solid-state materials-based sensors.

1.2 The definition of pH measurement

The pH value is used for monitoring the chemical characteristics of the materials in solid or liquid form. pH stands for the power of hydrogen, and it measures the

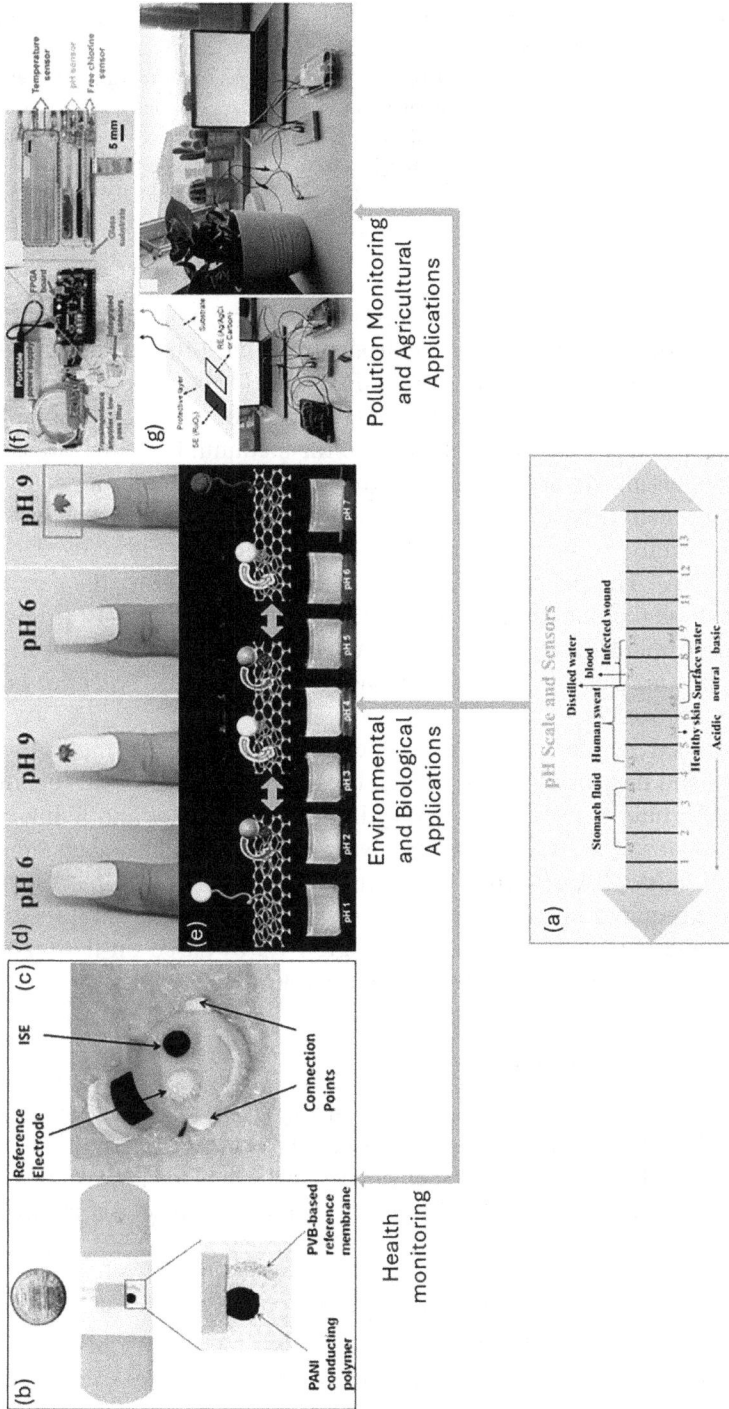

Figure 1.2. (a) pH scale and important values for various solutions. (b) Bandage-based wearable for monitoring wound pH [36], John Wiley & Sons. Copyright 2014 WILEY-VCH Verlag GmbH & Co.

KGaA, Weinheim. (c) Tattoo-based ion-sensitive electrodes for monitoring of epidermal pH levels, reproduced from [37] with permission from the Royal Society of Chemistry. (d) Wearable chemical pH sensor based on fingernail platform, reproduced from [38], copyright (2016), with permission from Elsevier. (e) Calorimetric graphene oxide-based pH sensor for environmental and biological applications, reprinted with permission from [39], copyright (2014) American Chemical Society. (f) Integrated water quality monitoring system measuring the pH value, reprinted from [35], copyright (2018), with permission from Elsevier. (g) pH sensor integrated with a circuit for precision agriculture, reprinted from [15], copyright (2023), with permission from Elsevier.

acidic, neutral and basic nature of the solution. The original definition of $pH = -\log[H^+]$ concerns the concentration of hydrogen ions $[H^+]$ in a solution. Since water dissociates into hydronium H_3O^+ and hydroxyl OH^- ions, the pH value describes the concentration of hydronium ions $[H_3O^+]$ in aqueous solutions. The pH value is normally measured at a given temperature of the medium. If the pH value is between 1 and 6, it represents an acidic solution, the pH value 7 denotes a neutral solution, and 8–14 represents a basic nature of the solution. The pH scale was introduced by the Danish Scientist Søren Peter Lauritz Sørensen. It covers the active concentration of H^+ and OH^-. The pH measurement of the solution is normally carried out by: (i) visual method (colour comparison of pH-sensitive litmus paper); (ii) photometric method (spectrophotometer to measure the wavelength of the pH-sensitive colored solution); and (iii) electrochemical approach (measuring the potential or electrical properties variation) [43–46].

As mentioned in the introduction, the pH value is very important in many applications and measuring the pH range is essential for their ideal function. When considering Nature, the pH value is very important in water and living organisms. The pH value of soil, water, plants, and living organisms defines their functions. An optimum pH value of soil is required for the ideal growth of a plant. As an example, the pH range required in soil for a few plants is shown in figure 1.3(a) [15]. The pH values for some food items are shown in figure 1.3(b) based on the reported work [43]. Values of body fluids are given in figure 1.3(c) [47].

1.3 Glass-based pH sensor

One of the gold standard methods for measuring the pH value of a solution via an electrochemical approach is the glass-based pH sensor, and the first model was introduced by F Haber and Z Klemensiewicz in 1909. This is developed based on the observation of potential difference across a thin glass membrane that separates two solutions with different pH values, and was obtained by M Cremer in 1906 [48]. The glass membrane is typically based on 72% $SiO_{2\%}$– 22% Na_2O–6% CaO (or 80%$SiO_{2\%}$–10%Li_2O–10%CaO for highly alkaline media) [43]. The glass electrode is fabricated based on pH-sensitive glass membrane (approximately 0.1 mm thickness) with an internal solution containing 0.1 N HCl', as shown in figure 1.4. The details of reference electrode fabrication are provided chapter 3. When this membrane comes in contact with the

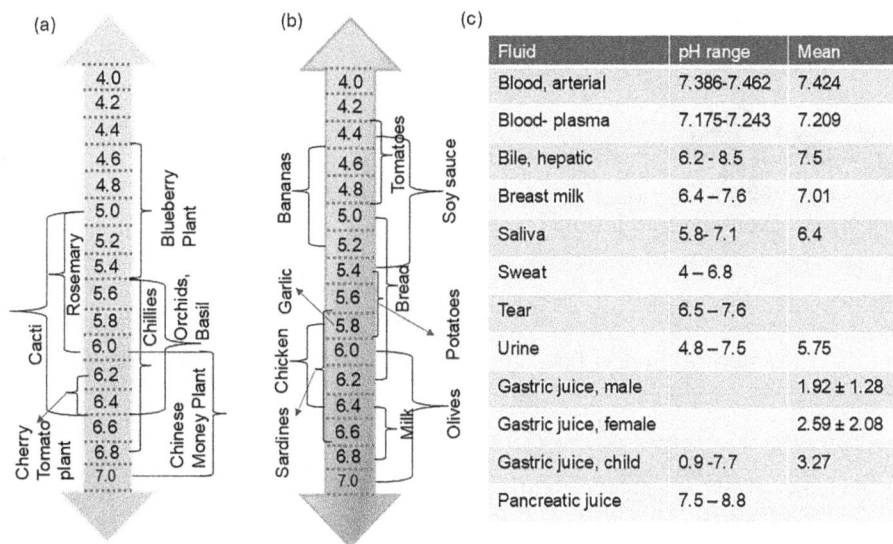

Figure 1.3. Important pH scale for (a) plants [15], (b) food [43] and (c) healthcare [47].

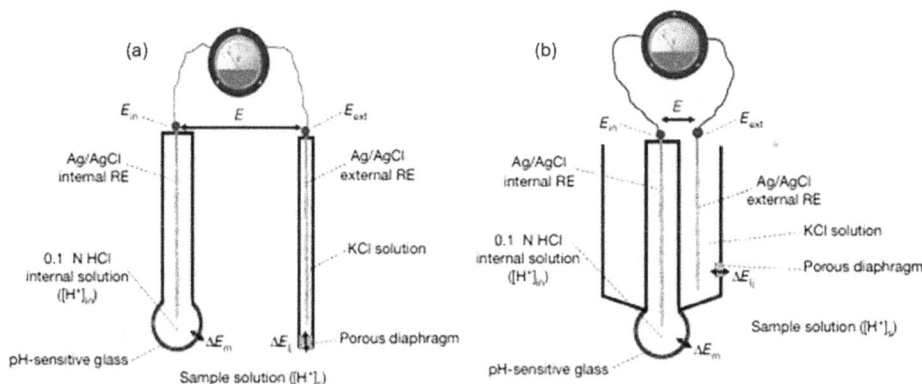

Fluid	pH range	Mean
Blood, arterial	7.386-7.462	7.424
Blood- plasma	7.175-7.243	7.209
Bile, hepatic	6.2 - 8.5	7.5
Breast milk	6.4 – 7.6	7.01
Saliva	5.8- 7.1	6.4
Sweat	4 – 6.8	
Tear	6.5 – 7.6	
Urine	4.8 – 7.5	5.75
Gastric juice, male		1.92 ± 1.28
Gastric juice, female		2.59 ± 2.08
Gastric juice, child	0.9 -7.7	3.27
Pancreatic juice	7.5 – 8.8	

Figure 1.4. Schematic representation of (a) glass pH-sensitive electrode with separate reference electrode and (b) in-built glass pH-sensitive and reference electrode. Reprinted from [43], copyright (2016), with permission from Elsevier. .

pH solution, it is hydrated and dissociates to Si–O$^-$ and this in turn is partially protonated.

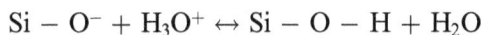

$$Si - O^- + H_3O^+ \leftrightarrow Si - O - H + H_2O$$

The potential difference or electromotive force is measured between two electrodes, such as a glass sensitive electrode and a reference electrode. These two combined are known as a pH meter, in which either both electrodes are separately connected or made as a single system, as shown in figures 1.4(a) and (b) [43].

Here, the measured potential between the terminals (E) is

$$E = E_{in} - \Delta E_m - \Delta E_{lj} - E_{ext}$$

E_{in} and E_{ext} are the potentials of the reference electrodes, ΔE_m is the drop between the two sides of the glass membrane and ΔE_{lj} is the liquid junction potential drop of the electrolytic junction. E_{in}, E_{ext} and ΔE_{lj} are constant. The potential drop ΔE_m is given by the Nernstian equation

$$\Delta E_m = 2.303 \frac{RT}{F} \log \frac{[H^+]_{in}}{[H^+]_s}$$

Here $[H^+]_{in}$ is a constant and hence the potential difference for the sample solution

$$E = E'_{const} - 2.303 \frac{RT}{F} pH_s$$

The glass-based pH sensing is highly selective to hydrogen ions due to its excellent sensitivity. In which the H^+ ions bind to the vacant oxygen ions in the glass membrane. These bindings of ions cause potential at the glass membrane solution interface, and they vary with changes in the concentration of ions. The potential difference between the glass membrane and the reference electrode depends on the concentration of ions, as schematically represented in figure 1.5(a) [49].

The major issues of a glass-based sensor are its lack of flexibility and miniaturization, and brittleness. The major research now focuses on the development of glass sensing membranes and new flexible designs in electrodes [49, 50]. One of the excellent designs reported for glass pH-sensitive electrodes is a flexible pH sensor fabrication. Figure 1.5(b) shows a novel design on a flexible printed glass-based pH-sensitive electrode. Here, the carbon electrode replaces the internal solution and acts as a conductive medium. The fabrication steps of the flexible glass-based pH sensor are shown in figure 1.6(a) and the developed sensor in figure 1.6(b) [49]. The developed sensor exhibited a sensitivity of 48 mV/pH in the range of pH 4–10. The major fabrication steps of the glass-based flexible pH sensor consist of [49]:

(i) Screen printing of the carbon-based electrode on a flexible polyethene terephthalate (PET) substrate. The printed electrodes were cured in an oven at 100 °C.

(ii) Removal of the contaminant on the electrode surface by oxygen plasma to clean and functionalize the surface. This plasma removes the contaminant in the form of water vapour and CO_2.

(iii) The precursor (2,4,6,8-tetramethylcyclotetrasiloxane) is supplied through the plasma head, where it reacts with activated oxygen and gets deposited in the form of silica on the carbon electrode.

(iv) Through hot pressing, the annealing of silica coating resulting in a higher degree of crosslinking through the formation of a higher number of Si–O–Si groups.

(v) Finally, the middle part of the sensor is passivated with a silicon glue to prevent any electrical short during the pH measurement in solution.

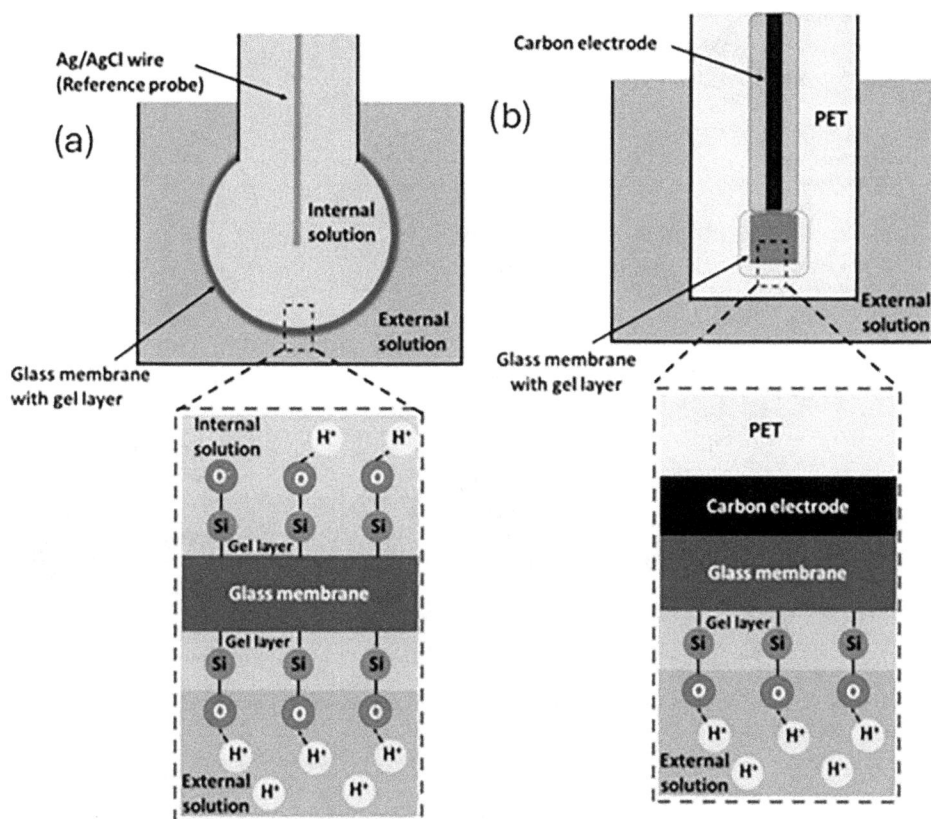

Figure 1.5. (a) pH sensing mechanism of conventional glass-based sensor, (b) pH sensing mechanism of the flexible glass-based pH sensing, (c) fabrication steps of the flexible glass-based pH sensor, and (d) image of the developed flexible glass-based pH sensor. Reprinted with permission from [49], copyright (2022) American Chemical Society.

1.4 Summary

Among various electrochemical and biosensor developments, the pH sensor received significant attention due to its importance in controlling the chemical reactions and compositions in various biochemical, chemical and biological processes. The pH measurement in the range of 0–14 is important in many fields. This chapter provides a summary of the importance of pH measurement, the range of pH values in important fields, and the progress of glass-based pH sensors. There is significant demand for new sensor design and materials, which are discussed in the following chapters. The technological advances in the past few decades have led to the development of new pH sensors. So far, various methods have been employed for measuring the pH of a solution, including both electrochemical and non-electrochemical methods. This chapter provides an overview of the definition, mechanism, and importance of electrochemical pH sensors.

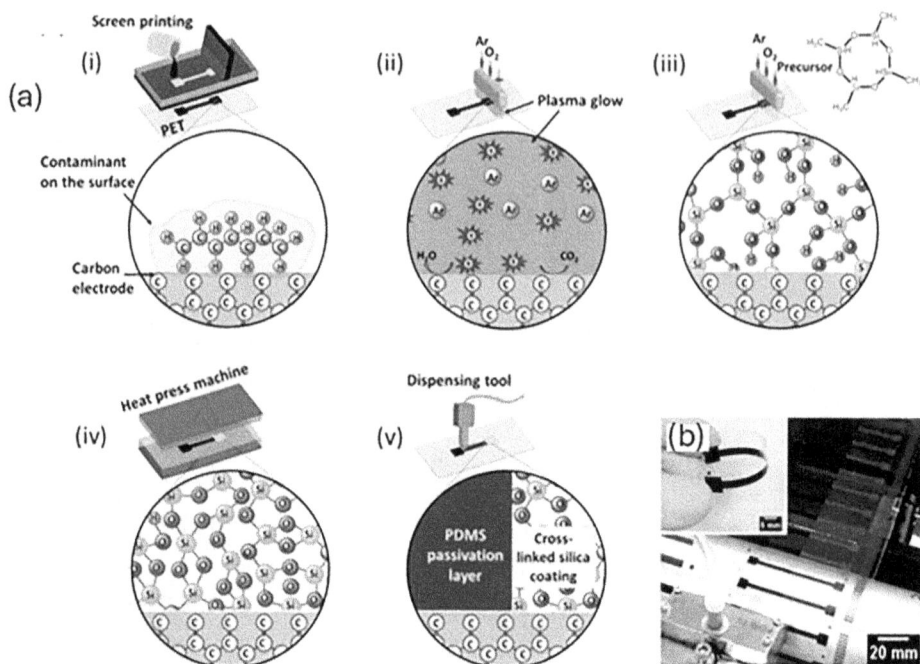

Figure 1.6. (a) Fabrication steps of the flexible glass-based pH sensor, and (b) image of the developed flexible glass-based pH sensor [49].

References

[1] Dissanayake M, Ramchandran L, Donkor O N and Vasiljevic T 2013 Denaturation of whey proteins as a function of heat, pH and protein concentration *Int. Dairy J.* **31** 93–9

[2] Law A J R and Leaver J 2000 Effect of pH on the thermal denaturation of whey proteins in milk *J. Agric. Food Chem.* **48** 672–9

[3] Andrés-Bello A, Barreto-Palacios V, García-Segovia P, Mir-Bel J and Martínez-Monzó J 2013 Effect of pH on color and texture of food products *Food Eng. Rev.* **5** 158–70

[4] Britten M and Giroux H J 2001 Acid-induced gelation of whey protein polymers: effects of pH and calcium concentration during polymerization *Food Hydrocoll.* **15** 609–17

[5] Wang X, Zhang N, Wu Z, Ma Y, Zhang H, Ma Y, Li H, Rayan A M, Ghamry M and Taha A M 2025 Construction of mixed gels of yolk granules and salted ovalbumin driven by pH: phase behavior and calcium bioaccessibility *Food Hydrocoll.* **159** 110682

[6] Zhang J, Shengmao L, Chenxue X, Chengyang W, Yi Z and Fan K 2024 Recent advances in pH-sensitive indicator films based on natural colorants for smart monitoring of food freshness: a review *Crit. Rev. Food Sci. Nutr.* **64** 12800–19

[7] Bansal V and Veena N 2024 Understanding the role of pH in cheese manufacturing: general aspects of cheese quality and safety *J. Food Sci. Technol.* **61** 16–26

[8] Leishangthem C, Mujumdar A, Xiao H W and Sutar P 2025 Intrinsic and extrinsic factors influencing *Bacillus cereus* spore inactivation in spices and herbs: thermal and non-thermal sterilization approaches *Compr. Rev. Food Sci. Food Saf.* **24** e70056

[9] Pinto C A, Mousakhani Ganjeh A, Barba F J and Saraiva J A 2024 Impact of pH and high-pressure pasteurization on the germination and development of *Clostridium perfringens* spores under hyperbaric storage versus refrigeration *Foods* **13** 1832

[10] Kochansky C J, McMasters D R, Lu P, Koeplinger K A, Kerr H H, Shou M and Korzekwa K R 2008 Impact of pH on plasma protein binding in equilibrium dialysis *Mol. Pharm.* **5** 438–48

[11] Kellum J A 2000 Determinants of blood pH in health and disease *Crit. Care* **4** 6

[12] Li X, Gong P, Zhao Q, Zhou X, Zhang Y and Zhao Y 2022 Plug-in optical fiber SPR biosensor for lung cancer gene detection with temperature and pH compensation *Sens. Actuators* B **359** 131596

[13] Manjakkal L, Cvejin K, Kulawik J, Zaraska K, Szwagierczak D and Socha R P 2014 Fabrication of thick film sensitive RuO_2–TiO_2 and Ag/AgCl/KCl reference electrodes and their application for pH measurements *Sens. Actuators* B **204** 57–67

[14] Uppuluri K, Szwagierczak D, Fernandes L, Zaraska K, Lange I, Synkiewicz-Musialska B and Manjakkal L 2023 A high-performance pH-sensitive electrode integrated with a multi-sensing probe for online water quality monitoring *J. Mater. Chem.* C **11** 15512–20

[15] Rabak A, Uppuluri K, Franco F F, Kumar N, Georgiev V P, Gauchotte-Lindsay C, Smith C, Hogg R A and Manjakkal L 2023 Sensor system for precision agriculture smart watering can *Results Eng.* **19** 101297

[16] Manjakkal L, Szwagierczak D and Dahiya R 2020 Metal oxides based electrochemical pH sensors: current progress and future perspectives *Prog. Mater Sci.* **109** 100635

[17] Ghoneim M, Nguyen A, Dereje N, Huang J, Moore G, Murzynowski P and Dagdeviren C 2019 Recent progress in electrochemical pH-sensing materials and configurations for biomedical applications *Chem. Rev.* **119** 5248–97

[18] Vonau W and Guth U 2006 pH monitoring: a review *J. Solid State Electrochem.* **10** 746–52

[19] Lin J 2000 Recent development and applications of optical and fiber-optic pH sensors *TrAC, Trends Anal. Chem.* **19** 541–52

[20] Choi M G, Han J M, Lim H, Ahn S and Chang S-K 2025 Colorimetric pH-sensing of artificial gastric fluid using naphthalimide-based CH acids *Spectrochim. Acta* A **326** 125166

[21] Li Z, Wang X, Fu X, Liu J, Liu Y and Zhang H 2025 Ratiometric fluorescent capillary sensor for real-time dual-monitoring of pH and O_2 fluctuation *Spectrochim. Acta* A **327** 125388

[22] Li Y, Song S, Song J, Gong R and Abbas G 2025 Electrochemical pH sensor incorporated wearables for state-of-the-art wound care *ACS Sens.* **10** 1690–708

[23] Manjakkal L, Dervin S and Dahiya R 2020 Flexible potentiometric pH sensors for wearable systems *RSC Adv.* **10** 8594–617

[24] Sinha S and Pal T 2022 A comprehensive review of FET-based pH sensors: materials, fabrication technologies, and modeling *Electrochem. Sci. Adv.* **2** e2100147

[25] Qin Y, Kwon H-J, Howlader M M and Deen M J 2015 Microfabricated electrochemical pH and free chlorine sensors for water quality monitoring: recent advances and research challenges *RSC Adv.* **5** 69086–109

[26] Khan M I, Mukherjee K, Shoukat R and Dong H 2017 A review on pH sensitive materials for sensors and detection methods *Microsyst. Technol.* **23** 4391–404

[27] Adib M R, Barrett C, O'Sullivan S, Flynn A, McFadden M, Kennedy E and O'Riordan A 2025 In situ pH-controlled electrochemical sensors for glucose and pH detection in calf saliva *Biosens. Bioelectron.* **275** 117234

[28] Rich A M, Rubin W, Rickli S, Akhmetshina T, Cossu J, Berger L, Magno M, Nuss K M, Schaller B and Löffler J F 2025 Development of an implantable sensor system for *in vivo* strain, temperature, and pH monitoring: comparative evaluation of titanium and resorbable magnesium plates *Bioact. Mater.* **43** 603–18

[29] Rahman F, Ryan A, Bocchino A, Galvin P and Teixeira S R 2025 Microneedle-based electrochemical sensors for real-time pH and sodium monitoring in physiological environments *Sens. Bio-Sens. Res.* **48** 100777

[30] Li Z, Sun W, Shi Z, Cao Y, Wang Y, Lu D, Jiang M, Wang Z, Marty J L and Zhu Z 2025 Development of an osmosis-assisted hollow microneedle array integrated with dual-functional electrochemical sensor for urea and pH monitoring in interstitial fluid *Sens. Actuators* B **422** 136606

[31] Cai Q Y and Grimes C A 2000 A remote query magnetoelastic pH sensor *Sens. Actuators* B **71** 112–7

[32] Pang P, Gao X, Xiao X, Yang W, Cai Q and Yao S 2007 A wireless pH sensor using magnetoelasticity for measurement of body fluid acidity *Anal. Sci.* **23** 463–7

[33] Li J, Albri F, Maier R R, Shu W, Sun J, Hand D P and MacPherson W N 2015 A micromachined optical fiber cantilever as a miniaturized pH sensor *IEEE Sens. J.* **15** 7221–8

[34] Gonska J, Schelling C and Urban G 2009 Application of hydrogel-coated microcantilevers as sensing elements for pH *J. Micromech. Microeng.* **19** 127002

[35] Qin Y, Alam A U, Pan S, Howlader M M R, Ghosh R, Hu N-X, Jin H, Dong S, Chen C-H and Deen M J 2018 Integrated water quality monitoring system with pH, free chlorine, and temperature sensors *Sens. Actuators* B **255** 781–90

[36] Guinovart T, Valdés-Ramírez G, Windmiller J R, Andrade F J and Wang J 2014 Bandage-based wearable potentiometric sensor for monitoring wound pH *Electroanalysis* **26** 1345–53

[37] Bandodkar A J, Hung V W, Jia W, Valdés-Ramírez G, Windmiller J R, Martinez A G, Ramírez J, Chan G, Kerman K and Wang J 2013 Tattoo-based potentiometric ion-selective sensors for epidermal pH monitoring *Analyst* **138** 123–8

[38] Kim J, Cho T N, Valdés-Ramírez G and Wang J 2016 A wearable fingernail chemical sensing platform: pH sensing at your fingertips *Talanta* **150** 622–8

[39] Paek K, Yang H, Lee J, Park J and Kim B J 2014 Efficient colorimetric ph sensor based on responsive polymer–quantum dot integrated graphene oxide *ACS Nano* **8** 2848–56

[40] Graham D J, Jaselskis B and Moore C E 2013 Development of the glass electrode and the pH response *J. Chem. Educ.* **90** 345–51

[41] Doi K, Asano N and Kawano S 2020 Development of glass micro-electrodes for local electric field, electrical conductivity, and pH measurements *Sci. Rep.* **10** 4110

[42] Neupane S, Subedi V, Thapa K K, Yadav R J, Nakarmi K B, Gupta D K and Yadav A P 2022 An alternative pH sensor: graphene oxide-based electrochemical sensor *Emerg. Mater.* **5** 509–17

[43] Karastogianni S, Girousi S and Sotiropoulos S 2016 pH: principles and measurement *Encyclopedia of Food and Health* vol 4 (Academic) pp 333–8

[44] Camões M F 2009 The quality of pH measurements 100 years after its definition *Accredit. Qual. Assur.* **14** 521–3

[45] Bates R G 1948 Definitions of pH scales *Chem. Rev.* **42** 1–61

[46] Covington A K, Bates R and Durst R 1985 Definition of pH scales, standard reference values, measurement of pH and related terminology (Recommendations 1984) *Pure Appl. Chem.* **57** 531–42

[47] Gaohua L, Miao X and Dou L 2021 Crosstalk of physiological pH and chemical pKa under the umbrella of physiologically based pharmacokinetic modeling of drug absorption, distribution, metabolism, excretion, and toxicity *Expert Opin. Drug Metab. Toxicol.* **17** 1103–24

[48] Vivaldi F, Salvo P, Poma N, Bonini A, Biagini D, Del Noce L, Melai B, Lisi F and Francesco F D 2021 Recent advances in optical, electrochemical and field effect pH sensors *Chemosensors* **9** 33

[49] Kasi V, Sedaghat S, Alcaraz A M, Maruthamuthu M K, Heredia-Rivera U, Nejati S, Nguyen J and Rahimi R 2022 Low-cost flexible glass-based ph sensor via cold atmospheric plasma deposition *ACS Appl. Mater. Interfaces* **14** 9697–710

[50] Nakayama S, Onishi K, Asahi T, Aung Y L and Kuwata S 2009 Response characteristics of all-solid-state pH sensor using $Li_5YSi_4O_{12}$ glass *Ceram. Int.* **35** 3057–60

IOP Publishing

Advanced Electrochemical pH Sensing Technologies
Scientific fundamentals and applications
Libu Manjakkal

Chapter 2

Electrochemical pH sensors: classification of materials and their working mechanism

2.1 Introduction

An electrochemical pH sensor is generally regarded as a device that measures the potential difference between two electrodes when immersed in an acidic or alkaline solution. The two electrodes mainly consist of a pH-sensitive electrode, designed to be responsive to H^+ in the solution, and a reference electrode, which provides a stable and known potential compared to the pH-sensitive electrode. Electrochemical sensors are classified based on their working mechanism and fabrication method. In the primary type of pH sensor, the pH values are determined through electrochemical reactions involving H^+ or OH^- with a glass electrode or advanced functional materials, including metal oxides (MOx), polymers, and carbon composites [1–5]. Various materials have emerged as sensitive electrodes for electrochemical sensing. These electrodes are used as bulk, thick, or thin films, and nanostructures for electrode fabrication. Most reported electrochemical pH sensors operate on the potentiometric method, where the electromotive force (emf) is measured between the sensitive and reference electrode. However, due to the lack of miniaturised reference electrodes and sensors, a new approach involving the measurement of electrical properties such as resistance, conductance, or impedance has led to the development of novel pH sensors.

Electrochemical pH sensors have found potential applications in many fields, including biomedical, water pollution monitoring, agriculture, food quality monitoring, regulation of chemical reactions and materials synthesis. Due to this significance, the past few decades have witnessed a considerable effort by researchers all over the world on the development of suitable electrochemical sensors based on different analysis approaches. This chapter provides information about the materials for electrochemical pH sensors and different methods of analysis employed in pH sensors.

2.2 Materials for electrochemical pH sensing

Many materials and designs have been investigated for the development of sensitive electrodes for pH sensors. The most popular electrodes are based on the glass electrode [3, 6, 7], Metal/MOx [3, 6, 8–10], MOx, carbon/graphene and polymers [5]. A comparison of the advantages and disadvantages of pH sensors fabricated based on these sensitive electrodes is given in table 2.1.

Table 2.1. Comparison of the major materials used for electrochemical pH sensor fabrication. Reproduced from [1] CC BY 4.0.

Materials	Advantages	Disadvantages
Glass electrode [3, 6, 7, 11–13]	– Very good pH response (close to Nernstian behaviour). – High accuracy and excellent stability over a wide pH range. – Long lifetime (several years) and low cross-sensitivity to other dissolved ions. – Very useful for practical applications in laboratories.	– Large in size, difficult for miniaturisation and expensive. – Mechanical fragility. – Chemical instability in corrosive systems and strong alkaline and hydrofluoric acid solutions. – Performance depends on high pressure and temperature – Requires frequent calibrations. – Limitations in the food industry and for online monitoring.
Polymer [4, 6, 11, 12, 14]	– Very good flexibility of the sensor. – Very good applicability for pH biosensors. – Conducting polymers exhibit high conductivity and electroactivity. – Non-conductive polymers have a highly selective response and high impedance, which eliminates interference from other ions.	– Adhesion problems. – Low reproducibility, reusability and long lifetime. – Poor reliability due to defects such as pinholes in some films. – Poor mechanical and chemical stability.
Metal/metal oxide [3, 4, 6, 8, 9, 11, 12, 15]	– Small and rugged sensors. – Easy for miniaturisation. – Mechanically and chemically stable. – Applicable for low impedance measuring systems. – Relatively stable and insensitive to salt concentration. – Useful for measuring body fluids such as those from the gastrointestinal tract.	– Limited pH sensing range. – Sensitive to several redox agents and oxygen pressure above the test solution. – Systematic deviation from Nernstian behaviour. – Poor resolution, repeatability and stability for polycrystalline electrode. – Very sensitive to several ligands and components of standard buffers. – Not ideal for blood monitoring.

| Metal oxide [3, 4, 6, 8, 11–13, 15–27] | – Very good sensitivity, close to Nernstian response.
– Fast response and long lifetime.
– High accuracy and selectivity to H+ ions.
– Low interferences due to other ions.
– Very good reusability, reproducibility and repeatability.
– Very low hysteresis, drift and optical effects (in particular for thick film potentiometric sensors).
– Low cost of fabrication and maintenance.
– Very good mechanical and chemical stability.
– The sensor can be stored in atmospheric conditions for a long time.
– Exhibits very good sensing performance in high temperature, high pressure, strong alkaline/acidic conditions and in corrosive systems.
– Potentiometric sensor does not require a power supply for operation.
– Conductimetric/chemiresistor-based sensor does not require a reference electrode.
– Easy miniaturisation for flexible/wearable systems.
– Compatible for online monitoring applications and biomedical, food processing, pollution control and nuclear respiratory systems. | – Some materials exhibit slow response in basic solutions.
– Lack of a compatible reference electrode, especially for thin film (ISFET) based sensor.
– Some materials show large drift, hysteresis and optical effect (especially some ISFET-based sensors).
– Conductimetric/chemiresistor-based sensor shows slow response to H + ions and is influenced by solution conductivity.
– Some materials show low accuracy and resolution.
– Sensing performance strongly depends on the microstructural properties of the film.
– Some materials show super-Nernstian or sub-Nernstian response. |

The comparison studies suggest that glass-based pH sensors perform very well, but their limitations have prompted the development of new solid-state materials for sensing electrodes. For real-time monitoring of solutions and miniaturised applications, the MOx-based sensitive electrode demonstrates superior performance. Recently, significant research has focused on thin and thick MOx films due to their excellent properties and suitability for a wide range of applications. A comparison of the performance of glass-based sensitive electrodes with advanced functional material-based on thin and thick film pH-sensitive electrodes is shown in figure 2.1(a). Although glass-based electrodes exhibit high sensitivity, selectivity, longevity, and quick

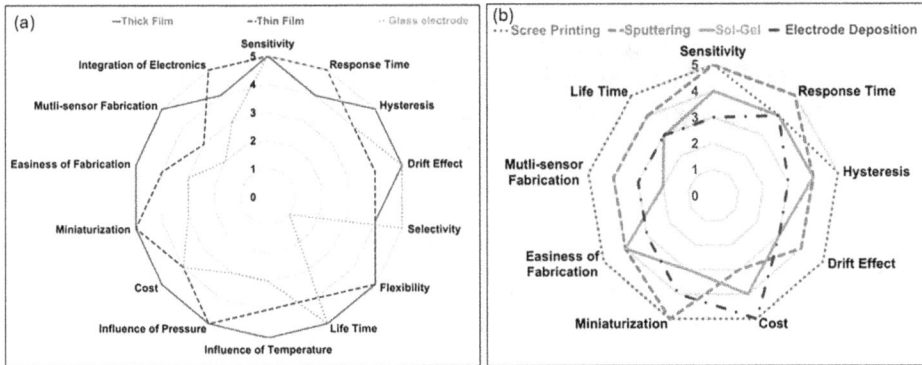

Figure 2.1. Comparison of (a) different types of electrodes, and (b) method of fabrication for pH sensing performances. Reproduced from [1] CC BY 4.0.

response times, thick film-based electrodes offer many advantages for the next generation of sensors. Many methods are implemented for the development of thin and thick film-based pH sensors. It was observed that the method of fabrication influences the parameters of performance, miniaturisation and cost of sensors. Figure 2.1(b) shows different approaches so far used for sensitive electrode development. In which the screen-printing method offers many advantages and has possibilities of developing multi-sensitive electrode and thick film reference electrode on the same substrate. The variation of performance of the sensors in various fabrication approaches could be due to: (i) the type of materials used; (ii) variation of thickness of the thin and thick film electrodes; and (iii) changes of the nano and microstructural properties of the electrodes. Based on these electrodes, various methods are considered for evaluating the sensing mechanism and performances of electrochemical pH sensors, and are discussed below.

2.3 Method of pH sensing performance analysis

2.3.1 Potentiometric analysis of the pH sensors

The pH in potentiometric-based sensors is determined by the potential difference between the reference electrode and sensing electrode when immersed in a solution of unknown pH. Potentiometric sensors are commonly used for measuring the pH of a solution using the Nernstian relation. The slope of the Nernstian response predicts the sensitivity of the sensor. A schematic representation of a screen-printed potentiometric pH sensor is shown in figure 2.2(a). Here, both the sensitive and reference electrodes are printed on the same substrate as shown in figure 2.2(b) and can be designed as flexible electrodes as well. During the electrochemical reaction, the electric charge distribution is time dependent. The Nernstian response of the sensor depends on the fabrication method of the electrode, the type of electrode materials kinetic and temperature properties of the measuring medium. The ideal Nernstian response is shown in figure 2.2(c). The potentiometric analysis provides the detailed performance of the pH sensor, including response time, stability,

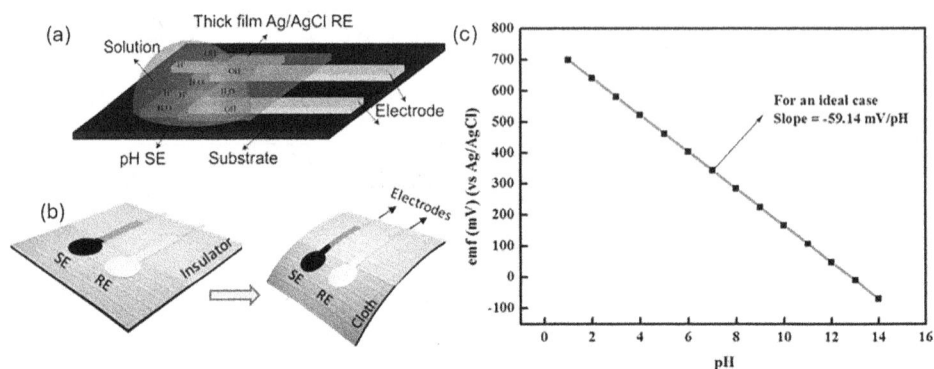

Figure 2.2. (a) Schematic representation of potentiometric pH sensor, reproduced from [1] CC BY 4.0. (b) Printing of sensitive electrode (SE) and reference electrode (RE) on flexible and non-flexible substrate, reproduced from [28] CC BY 4.0. (c) Ideal Nernstian response of a pH sensor, reproduced from [1] CC BY 4.0.

selectivity, interference of other ions, hysteresis and drift of a potentiometric thick film pH sensor.

2.3.1.1 Response time

In the potentiometric approach, the response of the sensors is measured by calculating the time (t_{90}) required for the electrode's open open-circuit potential to reach 90% of an equilibrium value [27]. In a thin or thick film-based pH sensor, the response time depends on the microstructure of the electrode, surface morphology, electrode thickness, pore size and temperature of the solution. In many works, it was found that in the acidic region, the sensor shows a faster response compared with the alkaline region. This is related to the diffusion of H^+ on the surface and bulk of the electrode. It was found that both the nanostructured nature and porosity of the sensing film may improve the response time of the sensor [27]. Further, the electrode fabrication method changes the surface morphologies and thickness of the electrodes, it leads to variation of the response time of the sensor [29]. The response times of a few sensors are summarised in table 2.2.

2.3.1.2 Hysteresis or memory effect in pH sensor

The differences in potential values when measuring at the same pH values multiple times are known as hysteresis or the memory effect. The representation of the measurement of hysteresis is shown in figure 2.3(a), and the voltage difference $\Delta V1$ and $\Delta V2$ represent the hysteresis for two pH values. In thin or thick film-based pH sensors, the hysteresis width depends on the measurement loop time and increases with increasing loop time [45]. Figure 2.2(b) shows the hysteresis curve of the SnO_2-based pH electrode, which depends on the loop time [46]. In addition, the hysteresis values are observed to be smaller in the acidic region as compared to the basic region of the pH solutions.

Hysteresis width is not only due to the material, but also because of the type of sensor, measurement method, and crystalline structure of the materials. A few

Table 2.2. Response time of the reported pH sensors.

Materials	Fabrication method	pH range	Response time	References
Pt-doped RuO$_2$	Screen printing	2–13	1–2 s	[30]
RuO$_2$–TiO$_2$	Screen printing	2–12	<15 s	[31]
RuO$_2$-commercial paste	Screen printing	1–12	<5 s	[17]
RuO$_2$ nanoparticles-MWCNT	Magnetron sputtering	2–12	<40 s	[27]
RuO$_2$–Ta$_2$O$_5$	Screen printing	2–12	<8 s and <15 s in acidic and basic solution	[32]
RuO$_2$–SnO$_2$	Screen printing	2–12	<5 and 9 s, for acidic to basic and basic to acidic solution changes	[33]
IrO$_2$ (flexible sensor)	Dip coating	1.5–12	0.9 s for pH 3.9–11; 2 s for pH 12–3.5	[25]
IrO$_2$–TiO$_2$ (30–70) mol%	Pechini method	1–13	120 s—pH 4–12; 10 s—acidic region; 5 s—basic region	[34]
IrO$_2$	Anodic electrodeposition	2–10	0.5, 1.5, and 1 min for pH regions <5, 5–7, and >7	[35]
Ta$_2$O$_5$	Sputtering	2–12	<0.3 s	[36]
TiO$_2$ nanotube array modified Ti electrode	Anodization of Ti substrate electrode	2–12	<30 s	[37]
WO$_3$/MWNT	Magnetron sputtering	2–12	<90 s	[38]
W/WO$_3$	Electro oxidation	2–12	For pH 6–7–3 s, for high pH—15 s	[39]
PbO$_2$	Electrodeposition	1.5–12.5	1 and <30 s in acidic medium and in alkaline solutions	[40]
PbO$_2$-based graphite epoxy	PbO$_2$–graphite epoxy composite was inserted into the glass tube	1–11	<15 s	[41]
MnO$_2$	Solid-state reaction	2–12	Few seconds in acidic and basic solutions	[42]
Cobalt oxide	Screen printing	1–12	<1 min	[43]
MnO$_2$/GPLE	Electrodeposition	1.5–12.5	20 s in acidic medium and 60 s in alkaline medium	[44]

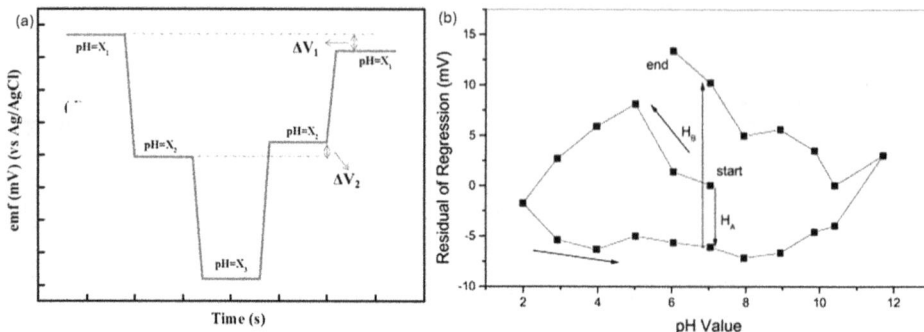

Figure 2.3. (a) Method measurement of hysteresis values of pH sensor. (b) The dependence of hysteresis width on the measurement loop time for SnO$_2$-based pH sensor, reprinted from [46], copyright (2005), with permission from Elsevier.

reports on hysteresis and new materials and methods indicate that the hysteresis effect is less in the case of MOx pH sensors. The observed hysteresis width of a few MOx-based pH sensors is given in table 2.3.

2.3.1.3 Interference effect of pH sensor

The interference effect or selectivity measurement is essential for pH sensing to distinguish the influence of other ions on the sensors. In solid-state materials, including MOx, polymers and carbon composite, also react with other dissolved ions in the water or any solutions. The influence of ions and their variation in the electromotive force are measured based on the Nikolsky–Eiseman equation, as given below as based on the International Union of Pure and Applied Chemistry (IUPAC) recommendation [52]

$$E = \text{constant} + \frac{2.303RT}{Z_A F} \log \left[a_A + K_{A,B}^{\text{pot}} a_B^{Z_A/Z_B} + K_{A,C}^{\text{pot}} a_C^{Z_A/Z_C} + \cdots \right]$$

Here, E is the measured potential in V, R is the universal gas constant (8.314 510 J K^{-1} mol^{-1}), T is the temperature in K, F is the Faraday constant (9.648 5309 × 104 C mol^{-1}), a_A is the activity of principal ion A, a_B and a_C are the interfering ions B and C, respectively. $K_{A,B}^{\text{pot}}$ potentiometric selectivity coefficient for ion B with respect to the principal ion A. Z_A is the charge number of the principal ion A. Z_B and Z_C are charge numbers of interfering ions B and C, respectively. Here the selectivity coefficient $K_{A,B}^{\text{pot}}$ is measured by two methods: fixed interference methods (FIM) and separate solution method (SSM). FIM is employed for mixed solutions of primary ion, A, and interfering or less desirable ion B [52]. The $K_{A,B}^{\text{pot}}$ is calculated using the Nikolsky–Eiseman equation [52]

$$K_{A,B}^{\text{pot}} = \frac{a_A}{a_B^{Z_A/Z_B}}$$

In the SSM, the selectivity coefficient is used to measure for separation of solutions, which contain ion activity A in one solution and the other consists of ion B. The selectivity coefficient measured from the measured potential values of E_A and E_B are

Table 2.3. Hysteresis effect of MOx-based sensors. Reproduced from [1] CC BY 4.0.

Materials	Method fabrication	pH loop/values	Hysteresis (mV)	References
RuO_2	RF sputtering	7–4–7–10–7	4.36	[26]
		7–10–7–4–7	2.2	
RuO_2	Magnetron sputtering	7–4–7–10–7	6.4	[27]
		7–10–7–4–7	5.1	
		2–8–12–8–2	10.2	
IrO_2	Sol–gel	1.50, 2.81, 3.75, 6.28, 7.86, 9.52 and 10.50	23.7, 9.5, 0.3, 14.5, 16.5, 6.7, and 11.5	[25]
IrO_2	Electrodeposition	2–10–2	2.5 ± 0.6	[35]
IrO_2	Electrodeposited	7	1.5–0.5	[47]
WO_3	Magnetron sputtering	2–12	<13	[38]
WO_3	RF sputtering	3–1–3–5–3	7.2	[48]
		5–3–1–3	12.5	
		4–1–4–7–4	12	
		4–7–4–1–4	26	
		3–1–3–5–3	7.2, 9.7 and 15.4 (10, 20 and 40 min)	
Ta_2O_5	RF sputtering	7–4–1–4–7–10–7	~5	[49]
Ta_2O_5	Electron-beam evaporation	3–10	1.5–9	[50]
Ta_2O_5		6–2–6–12–6	15.2 and 0.3 (without and with post-annealing)	
		6–12–6–2–6	6.3 and 0.7 (without and with post-annealing)	
SnO_2	RF sputtering	7–10–7–4–7	<3	[46]
		7–4–7–10–7	<7	
		7–2–7–12–6	7.3	
SnO_2	Sol–gel	5–1–5–9–5	3.74	[51]
		4–1–4–7–4	1.3	

$$\log K_{A,B}^{\text{pot}} = \frac{(E_B - E_A)z_A F}{2.303RT} + \left[1 - \frac{z_A}{z_B}\right]\lg a_A$$

If the ions a_A and a_B are giving same voltage in SSM, the selectivity coefficient measures

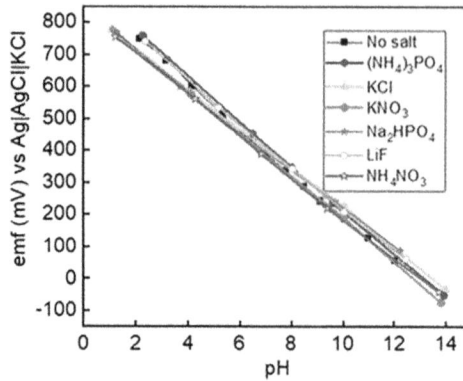

Figure 2.4. The measurement of influence of ions to the pH sensing. Reproduced from [53] based chemiresistive pH sensor.

Table 2.4. Measurement of selectivity coefficient for various ions for a Co_2O_3–RuO_2-based pH-sensitive electrode. Reproduced from [53] with permission from the Royal Society of Chemistry.

Interfering salts	Sensitivity (mV/pH)	E° (mV)	Linearity R^2	Calculated selectivity coefficient $k_{H^+/ion}$
No salts added	70.5	891	0.998	—
KCl	67.2	869	0.996	10^{-12}
KNO$_3$	66.8	850	0.999	10^{-11}
NH$_4$NO$_3$	69.8	865	0.997	10^{-11}
(NH$_4$)$_3$PO$_4$	68.6	884	0.996	10^{-11}
Na$_2$HPO$_4$	65.8	870	0.999	10^{-12}
LiF	69.0	902	0.998	10^{-11}

$$K_{A,B}^{\text{pot}} = \frac{a_A}{a^{z_A/z_B}}$$

In one of the case studies, for a Co_2O_3–RuO_2-based pH-sensitive electrode, the selectivity coefficient was measured by using the FIM method and figure 2.4 shows the variation of emf response with pH value of solutions for various ions. Considering hydrogen ion as primary ion and the other ions (Li^+, K^+, NH_4^+ and Na^+) as interfering or secondary ions, the selectivity coefficient was measured [53]. The negligible value, as shown in table 2.4 predicts that the sensitive electrode has less interference to other ions and is highly sensitive for the pH value change of the solution.

2.3.1.4 Drift effect of pH sensor
The non-random change of the output emf of the pH sensor with time in solution at constant temperature and composition is known as the drift effect. As compared to the glass-based pH sensors, in many metal oxide or solid-state materials-based pH

Table 2.5. Drift effect comparison of a few MOx-based sensors for different pH values of the solution. Reproduced from [1] CC BY 4.0.

Materials	Method of fabrication	Drift effect	References
RuO_2	Radio frequency sputtering	pH 4: 0.13 mV/pH; pH 7: 0.38 mV/pH; pH: 10 7.31 mV/pH	[26]
RuO_2	Screen printing	± 1.5 mV/month (after 15 days of measurement)	[30]
IrO_2	Electrodeposition	pH 1: 0.02 pH/h pH 7: 0.003 pH/h pH 11: 0.07 pH/h	[47]
IrO_2	Sol–gel	3–10 mV	[25]
Ta_2O_5	Radio frequency sputtering	pH 7: 0.03–0.05 mV/day	[36]
Ta_2O_5	Electron-beam evaporation	pH 7: <0.5 mV h^{-1} (after 3 h); 0.2 mV h^{-1} (after 10 h)	[50]
WO_3	Radio frequency sputtering	pH 1: 1.5 mV h^{-1} pH 3: 3.6 mV h^{-1} pH 5: 6.6 mV h^{-1} pH 7: 15.7 mV h^{-1}	[48]
WO_3	Magnetron sputtering	<3 mV	[38]
TiO_2	Sol–gel	pH 7: 1.97 mV h^{-1}	[55]
TiO_2	Radio frequency sputtering	1.67 mV h^{-1}	[56]

sensors, the potential drift is found to be larger. The drift effect depends on the H^+ transport in the material's surface or in the bulk, and the drift rate increases with an increase in time. The previous studies observed that (i) the OH^- is one of the factors that causes higher drift in alkaline solutions than in acidic solutions, (ii) the temperature of the solution medium causes the variation in drift, (iii) the method of fabrication of electrodes and their compositions also influence the drift of the emf [1, 30, 45, 54]. The comparison of the drift effect of a few metal oxides is shown in table 2.5 [1].

2.3.2 Chemiresistive analysis of the pH-sensitive electrode

One of the simple methods of fabrication and analysis of pH sensing performances is to use a chemiresistive approach. In a chemiresistive pH sensor, the sensitive electrode is deposited between a conductive electrode. The chemiresistive approach is implemented to overcome the issues of a miniaturised reference electrode. In a chemiresistive pH sensor, the sensing properties are analyzed based on changes in electrical properties such as conductance, capacitance, resistance or impedance of a film deposited between two electrodes or interdigitated electrodes as a result of the electrochemical reaction at the solution-sensitive electrode interface. The changes in H_3O^+/OH^- in the solution generate a change in the electrical properties of the sensitive electrode, especially resistance and conductance of the material. The schematic representation of the chemiresistive sensor is shown in figure 2.5(a). For a chemiresistive pH sensor fabrication, MOx nanoparticle (e.g., TiO_2, CuO), conductive polymers and carbon/graphene-based electrodes are widely employed [57, 58]. Figure 2.5(b) represents the change in conductance of a TiO_2 nanowire (NW)-based chemiresistive pH sensor [59]. It was found that with increasing pH from acidic to basic, the conductance value decreases linearly with a resolution of 5.68 ± 0.28 nS/pH. The decrease in conductance is due to the reduction and increase in depletion layer at the surface of the TiO_2 NWs, which react with H^+/OH^- ions [59].

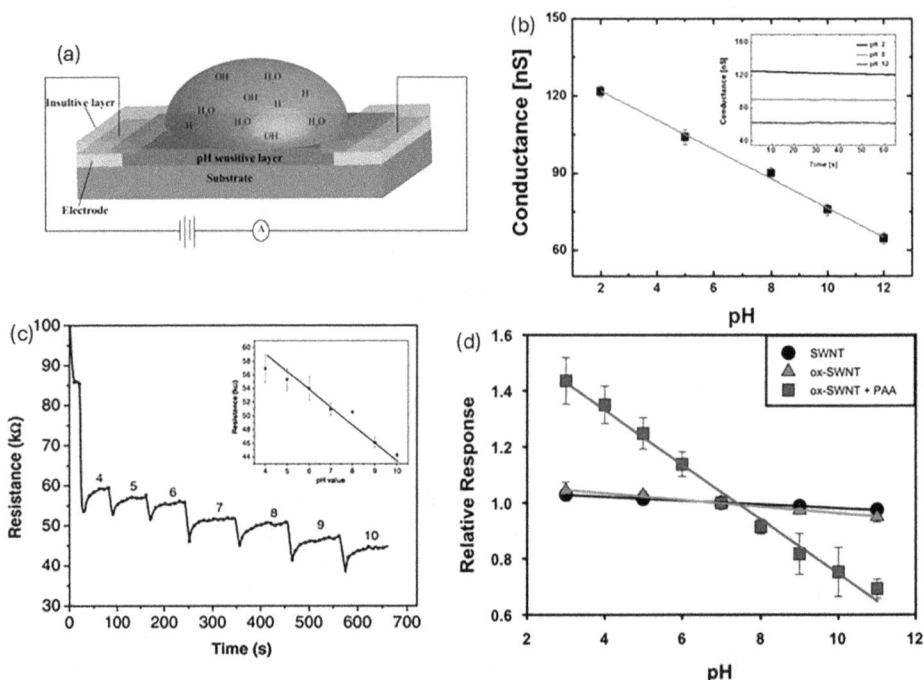

Figure 2.5. (a) Schematic representation of a chemiresistor-based pH sensor, reproduced from [1] CC BY 4.0. (b) Variation of conductance of TiO_2 NWs–TiO_2/C NF based pH sensor with an inset representing the conductance versus time at pH values of 2, 8, and 12, reprinted with permission from [59], copyright (2014) American Chemical Society. (c) Real-time resistance measurements of the graphene-based chemiresistor. Inset: resistance vs pH determined from multiple measurements, reproduced from [22], copyright IOP Publishing Ltd. All rights reserved. (d) Comparison of relative response for SWNTs ox-SWNT and ox-SWNT + PAA (ox-SWNT/PAA) used as conductance pH sensors, reproduced from [60] CC BY 4.0.

There are a few reports on chemiresistor-based pH sensors based on graphene and CNTs [22, 60]. Figure 2.5(c) represents the real-time resistance measurements of a graphene-based chemiresistor pH sensor at a current 10 μA [22]. The variation of resistance with pH of solution is shown in the inset of figure 2.3(c) [22]. A chemiresistor wireless pH sensor based on CNTs was reported by Gou et al [60]. They also investigated the sensing performance of a chemitransistor-based pH sensor utilising the same material [60]. Gou et al [60] compared the pH sensing performance of single-walled carbon nanotubes (SWNTs) with oxidised SWNTs (ox-SWNTs) and ox-SWNTs functionalized with the conductive polymer poly(1-aminoanthracene) (PAA) (ox-SWNT + PAA), as presented in figure 2.5(d). It was observed that ox-SWNT + PAA exhibited better sensing performance than ox-SWNT. This is due to the presence of carboxylic groups on the ox-SWNT. In the pH sensing performance, the oxygen-containing groups on the surface of SWNTs most likely play a role in pH response. Here, the conductive polymer PAA contributes to the device selectivity, while the CNTs provide a sensitive and robust platform necessary to chemically stabilise the polymer [60].

2.3.3 Electrochemical impedance spectroscopic analysis

Electrochemical impedance spectroscopy (EIS) analysis is used to investigate the electrode-electrolyte interaction of the pH sensor, and is one of the best approaches used for analysing the electrochemical reaction of the material. EIS enables us to predict the sensing mechanism of a solid-state material-based pH sensor by getting insight into phenomena such as adsorption, ion exchange, charge transfer, diffusion, etc. EIS analysis is normally employed for three-electrode measurements and two-electrode measurements of a sensor. In a three-electrode configuration, the experimental electrochemical cell setup consists of a working electrode, a counter electrode and a reference electrode [61]. In a three-electrode configuration, the solution resistance is compensated between the counter and reference electrode. One of the best designs of a sensor with a two-electrode configuration for EIS analysis is the structure with an IDE [19, 61, 62]. Moreover, another type of two-electrode configuration comprises a working (serving as a combined working and reference electrode) and a counter electrode (serving as a combined counter and second reference electrode) in the electrochemical cell setup [61]. In this case, the measured impedance of the sensor consists of the total impedance of the counter electrode, the working electrode and the sample [61]. In this electrochemical cell setup for minimising the influence of the counter electrode, a larger surface area of the counter electrode than the working electrode is employed for the sensor measurement [61].

In EIS, the impedance data is expressed in Nyquist and Bode plots, which provide information about the sensing mechanism [31, 33, 62–64]. These plots are used to analyse conventional electric circuits and the equivalent circuits to describe electrochemical properties of the electrodes and their interaction with a solution. There are some limitations for the circuit analysis by Nyquist plot, as the capacitive features of the electrode material may cause undesired variations. For these reasons, the Bode plot is often used to get more detailed information about impedance concerning the applied frequency. The Bode plot reveals the capacitive, resistive and inductive features of the electrochemical system. In most of the electrode-solution interface reactions, the variation in impedance of a Faradaic reaction may be due to: (1) adsorption of reacting species; (2) diffusion of ions; and (3) both diffusion and adsorption processes [65]. These phenomena may be related to the contribution of the electrolyte, the electrode-solution interface, the nature of the material and the electrochemical reaction occurring on the electrode. Here, we discuss the EIS analysis of two-electrode and three-electrode electrochemical systems.

Chen *et al* [49] observed that for a Ta_2O_5-based electrolyte-ion-sensitive membrane-oxide-semiconductor (EIOS) pH sensor, the membrane and electrolyte interface leads to a network consisting of the double layer capacitance (C_{dl}) and charge transfer resistance (R_{ct}). In this study, a two-electrode configuration (SE—Ta_2O_5 and RE—conventional Ag/AgCl) was used for electrochemical analysis of the sensor. In the lower frequency range, the sensor showed Warburg impedance due to diffusion of ions from the electrolyte to the membrane, as illustrated in figure 2.6(a) [49]. The authors evaluated the electrochemical parameters of the system by using ZsimpWin software [49]. EIS analysis of a sensor based on RuO_2/MWCNT by using a three-electrode configuration (SE—RuO_2/MWCNT, RE—Ag/AgCl (3 M KCl) and CE—platinum wire) revealed that R_{ct} increases gradually from 1.8 to 10.50 kΩ within the pH range 2–12 [27]. The impedance spectra of the sensor at pH 8 and its equivalent circuit is shown in figure 2.6(b).

A planar interdigitated electrode (IDE)-based pH sensor is a simple two-electrode probe configuration of an electrochemical cell setup and is widely used for miniaturised applications. The electrochemical reaction at the sensitive electrode-solution interface of an IDE-based pH sensor can be investigated by using EIS analysis. The EIS analysis of an IDE-based pH sensor indicates that the sensing mechanism of thick film pH sensors can be attributed to a combination of electron transfer and ion exchange processes at the solution/MO_x interface, as illustrated in figure 2.6(c). In the high-frequency range, the conductivity (R_s-solution resistance) and charge transfer processes (R_{ct}—charge transfer resistance) depend on the solution and sensing layer properties are dominant. However, in the low-frequency range, ion exchange (adsorption/diffusion) processes prevail, and sensor characteristics strongly depend on the solution pH. Thus, the low-frequency range is more interesting for the pH sensing purpose. The ion exchange processes also depend on the material properties like structure, composition, porosity and surface homogeneity [23, 33, 62, 64]. The Nyquist diagram for a thick film pH sensor based on RuO_2 paste (commercial paste) in the frequency range of 10 Hz–2 MHz is displayed in figure 2.6(d). The plot consists of two arcs. The diameter of the smaller semicircle is slightly changing with pH values, whereas the bigger semicircle is strongly pH dependent. However, in the low-frequency range (10 mHz–100 kHz) the arc is complete, as shown in inset (right side) of figure 2.6(d) [62].

In Bode plots, shown in the insets of figure 2.6(d) (right side), in the low-frequency range, the phase angle assumes significantly negative values, indicating that the impedance becomes partially capacitive in this range due to ionic exchange.

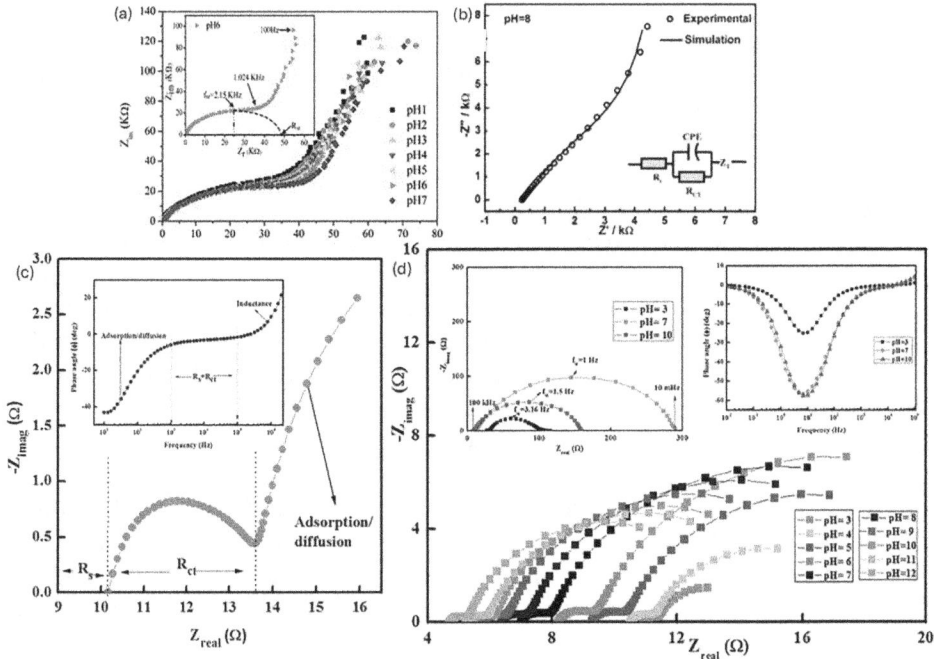

Figure 2.6. (a) EIS spectra of Ta$_2$O$_5$-based EIOS pH sensor, reprinted from [49], copyright (2014), with permission from Elsevier. (b) EIS of the RuO$_2$/MWCNTs electrode in solution of pH = 8 and its equivalent circuit (inset), reprinted from [27], copyright (2010), with permission from Elsevier. (c) Nyquist plot and Bode plot (inset) of RuO$_2$-based thick film pH sensor. reprinted from [64], copyright (2016), with permission from Elsevier. (d) EIS analysis of RuO$_2$ (commercial paste) based pH sensor: Nyquist plot (10 Hz–2 MHz), left side inset -Nyquist plot (10 mHz–100 kHz) and right side inset—Bode plot, reprinted from [62], copyright (2015), with permission from Elsevier.

However, in the intermediate frequency range (10^3–10^5 Hz), the phase angle is close to zero degrees, representing the purely resistive nature of the circuit. This may be due to the dominant influence of the R_s and the R_{ct} of the sensor in this frequency range [61, 62, 66]. Furthermore, at high frequencies >10^5 Hz the phase angle is shifted to positive values. It may be caused by the influence of the inductance of the sensor and the connecting wires on the impedance spectra [33, 61, 62, 64]. The parameters R_{ct}, C_{dl} (double layer capacitance), R_a and C_a (resistance and capacitance dependent on the activation energy of adsorption/desorption processes of ions at the metal oxide surface) and R_s which were derived from the complex impedance plots, are investigated by using MEISP (Multiple EIS Parameterization) software [62].

2.3.4 Amperometric analysis

In amperometric pH sensing external potential is applied to the sensitive electrode for the counter and reference electrode which results in redox reaction involving proton transfer. The generated amperometric curves vary with the pH value of the solution and are based on the shift of pH-sensitive peak potential of the redox species [67]. In an amperometric pH sensor, the pH sensitivity is also measured based on the measurement of current variation due to the redox reactions. For amperometric pH sensing, many redox-active mediators are employed for the fabrication. In one of the works, haematein on a *n*-eicosane–graphite composite electrode as the pH-sensitive redox-active mediator was employed, and it shows in different pH solutions the anodic and cathodic peaks shifts, as shown in figure 2.7(a) [68]. The variation of current with pH of solution for the same electrode is shown in figure 2.8(b) [68]. Amperometric analysis is mainly implemented for measuring the pH sensing performance of organic materials. For example, in conductive polymer PANI, at different pH values, the sensitive film will be protonated or deprotonated. This reaction leads to varying the charge density at the interface of PANI/pH solution [67]. This pH-dependent interface potential leads to a shift in the cyclic voltammogram [67].

In a graphene oxide-based sensor, the amperometric analysis is carried out to understand the electrochemical performances of the graphene oxide surfaces for heterogeneous electron transfer (HER) ability [69]. The HER depends on the extent of oxidation and surface functional groups (–OH, –COOH etc). The electrochemical setup for the amperometric measurement of the sensing electrode is shown in figure 2.7(c) [69].

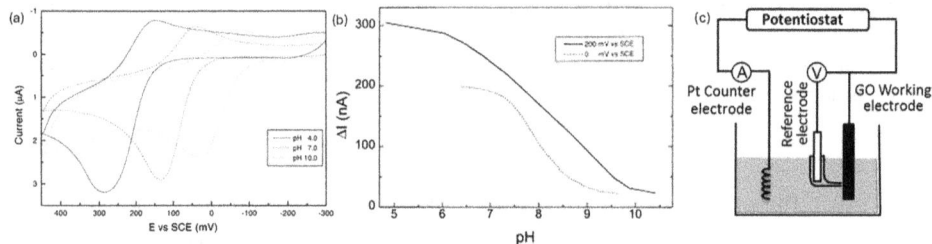

Figure 2.7. (a) Amperometric analysis for haematein on a n-eicosane–graphite composite electrode in different pH solution, and (b) its observed current changes with pH values. Reprinted from [68], copyright (2001), with permission from Elsevier. (c) A schematic representation of an amperometric measurement setup, reproduced from [69] with permission from Springer Nature.

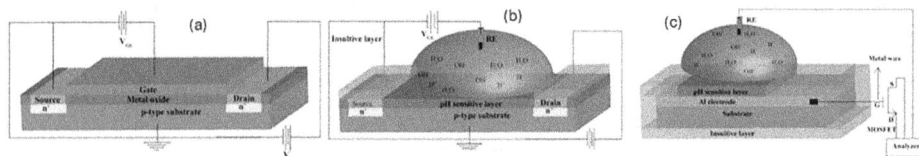

Figure 2.8. Schematic representation of (a) MOSFET, (b) ISFET and (c) EGFET with measurement setup. (b) and (c) reproduced from [1] CC BY 4.0.

2.3.5 ISFET-based pH sensor

The ion-sensitive field-effect transistor (ISFET) was introduced long ago as a iniaturized electrochemical pH and biosensor [20, 70, 71]. It is a type of potentiometric device which operates similar to metal oxide field-effect semiconductor transistor (MOSFET) [71] (shown in figure 2.8(a)). In 1970, Bergveld [20] used the first ISFET for measurements of ion activities in electrochemical and biological environments. In ISFET, the gate is covered with an ion-sensitive layer and is placed between the source and drain (figure 2.8(b)). The current flowing between source and drain is controlled by the electrostatic field generated by the gate. In this method, instead of the potential difference on two sides of the glass membrane the current passing through the transistor is observed. A wider range of materials used for making ion-sensitive layers on a gate include ceramics, organics, polymers or catalytic metal layers [12, 20, 36, 70–72]. The pH sensing mechanism of the oxide surface of a gate electrode can be described by the site-binding theory [12, 73]. Detailed investigations of ISFETs as pH and biosensors are described in numerous review articles including [74–76].

The optical effect, hysteresis, and drift are the three non-ideal characteristics observed in ISFET-based pH sensors. A detailed discussion about drift effect and hysteresis effect of MOx pH sensors were given in a previous section. Optical effect is an outer non-ideal characteristic, while hysteresis and drift are considered as inner non-ideal characteristics [6, 26]. This effect was first reported by Voorthuyzen and Bergveld [77] for Ta_2O_5 gate ISFET. Liao and Chou [26] found that an RuO_2-based ISFET device showed the variation of output voltage to be about 16.8 mV when measured under light exposure at pH = 7.

ISFETs can be easily adapted to a broad range of chemical, biomedical and biochemical measurements [40, 41] and compared to glass pH sensors, they have many advantages, including miniaturised size, low cost and fast response. Since ISFETs are directly formed on the FET electrode, they also have several disadvantages such as device instability, low current sensitivity and sensitivity to light etc [12]. To overcome these drawbacks, van der spiegel *et al* [78] Introduced extended gate field-effect transistor (EGFET), where the FET is isolated from the chemical solution as the sensitive film is deposited at the end of the signal line extended from the FET electrode [79–81], as illustrated in figure 2.8(c). EGFETs are widely used for biosensor applications [21, 78–82], and their major advantages are long-term stability, insensitivity to light, temperature drift, and suitability for disposability. The pH sensing mechanism of EGFET can be described by the site-binding theory. Only a few semiconductor MOx were utilized for EGFET-based pH sensor fabrication. As an example, the sensing performance transfer characteristics of a sensor based on SnO_2 deposited on FTO (fluorine-doped tin oxide) glass substrate show that the sensitivity of SnO_2 nanorod is (55.18 mV/pH) higher than SnO_2 thin film (48.04 mV/pH) [83]. A nanorod-based sensor showed high sensitivity due to the high surface-to-volume ratio for the nanostructure of the material, which provided larger surface sites and effective sensing areas.

In addition to ISFET and EGFET, recently, there have been many reports on the fabrication of chemi-transistors [84, 85], electrolyte-gated field-effect transistor (ElGFET), reparative extended gate ion-sensitive field-effect transistor (SEGISFET)-based pH sensors [86]. Organic semiconductors are mainly used for the development of chemi-transistors and ElGFETs [12] based pH sensors. Qin *et al* [12] recently reported a detailed comparative study of the structure and operating mechanism of such types of microfabricated electrochemical pH sensors.

Organic thin film transistors (OTFTs) based on organic semiconductors showed promising applications in biosensors and chemical sensors. The major advantages of such sensors are: (i) low temperature deposing of materials; (ii) good mechanical properties including flexibility; (iii) organic materials are compatible with glass, plastic, metal foils or polymer substrates; (iv) low cost of fabrication; (v) use as a disposable sensor; (vi) direct integration of devices with biological systems; and (vii) applicability in food processing etc [87, 88]. However, as compared to inorganic semiconductors, the organic semiconductor-based TFTs exhibited low mobility and environmental stability. The pH sensing applications of OTFT were first introduced by Bartic *et al* [89]. A detailed study on OTFTs for chemical and biological sensing applications was discussed by Lin and Yan [90]. Poly 3-hexylthiophene (P3HT), pentacene and polyaniline are the major organic semiconductors used for the fabrication of OTFT-based pH sensors. Loi *et al* [87] reported flexible organic ISFETs based on pentacene films on MylarTM sheet substrate for pH sensing. The pentacene films exhibited high hole mobility and good properties for flexible applications [87].

2.4 Summary

The advances in pH sensor technology are related to ways to improve sensitivity, reduce cost, enhance portability, and achieve rapid response in a wide range of pH values. Solid-state materials-based sensing electrodes are an ideal electrode for pH sensor applications, as confirmed in recent years. A systematic investigation is necessary for further improvement in the fabrication and application of new electrochemical pH sensors. Especially, it is very important to investigate the ionic exchange between the nanomaterial and electrolyte solution for electrochemical sensors and biosensors. The variation in ionic concentration of the solution, such as pH value, should significantly change the electrochemical properties, including electrochemical potential, impedance and capacitance of the electrodes. This chapter provides an overview of the classification of electrochemical sensors and measurement methods. Further detailed studies are required in reference electrode preparation for the completion of electrochemical pH sensors and are discussed in the next chapter.

Parts of this chapter have been reproduced from [1]. CC BY 4.0.

References

[1] Manjakkal L, Szwagierczak D and Dahiya R 2020 Metal oxides based electrochemical pH sensors: current progress and future perspectives *Prog. Mater Sci.* **109** 100635

[2] Ghoneim M, Nguyen A, Dereje N, Huang J, Moore G, Murzynowski P and Dagdeviren C 2019 Recent progress in electrochemical pH-sensing materials and configurations for biomedical applications *Chem. Rev.* **119** 5248–97

[3] Vonau W and Guth U 2006 pH monitoring: a review *J. Solid State Electrochem.* **10** 746–52

[4] Korostynska O, Arshak K, Gill E and Arshak A 2007 Review on state-of-the-art in polymer based pH sensors *Sensors* **7** 3027–42

[5] Salvo P, Melai B, Calisi N, Paoletti C, Bellagambi F, Kirchhain A, Trivella M G, Fuoco R and Di Francesco F 2018 Graphene-based devices for measuring pH *Sens. Actuators* B **256** 976–91

[6] Kurzweil P 2009 Metal oxides and ion-exchanging surfaces as pH sensors in liquids: state-of-the-art and outlook *Sensors* **9** 4955

[7] Eisenman G 1967 *Glass Electrodes for Hydrogen and Other Cations* (Marcel Dekker)

[8] Głab S, Hulanicki A, Edwall G and Ingman F 1989 Metal–metal oxide and metal oxide electrodes as pH sensors *Crit. Rev. Anal. Chem.* **21** 29–47

[9] Edwall G 1978 Improved antimony–antimony (III) oxide pH electrodes *Med. Biol. Eng. Comput.* **16** 661–9

[10] Horton B E, Schweitzer S, DeRouin A J and Ong K G 2011 A varactor-based, inductively coupled wireless pH sensor *IEEE Sens. J.* **11** 1061–6

[11] Yuqing M, Jianrong C and Keming F 2005 New technology for the detection of pH *J. Biochem. Biophys. Methods* **63** 1–9

[12] Qin Y, Kwon H-J, Howlader M M and Deen M J 2015 Microfabricated electrochemical pH and free chlorine sensors for water quality monitoring: recent advances and research challenges *RSC Adv.* **5** 69086–109

[13] Guth U, Vonau W and Zosel J 2009 Recent developments in electrochemical sensor application and technology—a review *Meas. Sci. Technol.* **20** 042002

[14] Antohe V A, Radu A, Mátéfi-Tempfli S and Piraux L 2011 Circuit modeling on polyaniline functionalized nanowire-templated micro-interdigital capacitors for pH sensing *IEEE Trans. Nanotechnol.* **10** 1314–20

[15] Fog A and Buck R P 1984 Electronic semiconducting oxides as pH sensors *Sens. Actuators* **5** 137–46

[16] Zhuiykov S 2012 Solid-state sensors monitoring parameters of water quality for the next generation of wireless sensor networks *Sens. Actuators* B **161** 1–20

[17] Martínez-Máñez R, Soto J, García-Breijo E, Gil L, Ibáñez J and Gadea E 2005 A multisensor in thick-film technology for water quality control *Sens. Actuators* A **120** 589–95

[18] Atkinson J K, Cranny A W J, Glasspool W V and Mihell J A 1999 An investigation of the performance characteristics and operational lifetimes of multi-element thick film sensor arrays used in the determination of water quality parameters *Sens. Actuators* B **54** 215–31

[19] Gill E, Arshak K, Arshak A and Korostynska O 2008 Mixed metal oxide films as pH sensing materials *Microsyst. Technol.* **14** 499–507

[20] Bergveld P 1970 Development of an ion-sensitive solid-state device for neurophysiological measurements *IEEE Trans. Biomed. Eng.* **1** 70–1

[21] Chang S-P and Yang T-H 2012 Sensing performance of EGFET pH sensors with CuO nanowires fabricated on glass substrate *Int. J. Electrochem. Sci.* **7** 5020–7

[22] Lei N, Li P, Xue W and Xu J 2011 Simple graphene chemiresistors as pH sensors: fabrication and characterization *Meas. Sci. Technol.* **22** 107002

[23] Manjakkal L, Synkiewicz B, Zaraska K, Cvejin K, Kulawik J and Szwagierczak D 2016 Development and characterization of miniaturized LTCC pH sensors with RuO_2-based sensing electrodes *Sens. Actuators* B **223** 641–9

[24] Wang J, Yokokawa M, Satake T and Suzuki H 2015 A micro IrO_x potentiometric sensor for direct determination of organophosphate pesticides *Sens. Actuators* B **220** 859–63

[25] Huang W-D, Cao H, Deb S, Chiao M and Chiao J-C 2011 A flexible pH sensor based on the iridium oxide sensing film *Sens. Actuators* A **169** 1–11

[26] Liao Y-H and Chou J-C 2008 Preparation and characteristics of ruthenium dioxide for pH array sensors with real-time measurement system *Sens. Actuators* B **128** 603–12

[27] Xu B and Zhang W-D 2010 Modification of vertically aligned carbon nanotubes with RuO_2 for a solid-state pH sensor *Electrochim. Acta* **55** 2859–64

[28] Manjakkal L, Dang W, Yogeswaran N and Dahiya R 2019 Textile-based potentiometric electrochemical pH sensor for wearable applications *Biosensors* **9** 14

[29] Zhuiykov S, Kats E, Kalantar-zadeh K, Breedon M and Miura N 2012 Influence of thickness of sub-micron Cu_2O-doped RuO_2 electrode on sensing performance of planar electrochemical pH sensors *Mater. Lett.* **75** 165–8

[30] Zhuiykov S 2009 Morphology of Pt-doped nanofabricated RuO_2 sensing electrodes and their properties in water quality monitoring sensors *Sens. Actuators* B **136** 248–56

[31] Manjakkal L, Cvejin K, Kulawik J, Zaraska K, Szwagierczak D and Socha R P 2014 Fabrication of thick film sensitive RuO_2–TiO_2 and Ag/AgCl/KCl reference electrodes and their application for pH measurements *Sens. Actuators* B **204** 57–67

[32] Manjakkal L, Zaraska K, Cvejin K, Kulawik J and Szwagierczak D 2016 Potentiometric RuO_2–Ta_2O_5 pH sensors fabricated using thick film and LTCC technologies *Talanta* **147** 233–40

[33] Manjakkal L, Cvejin K, Kulawik J, Zaraska K, Szwagierczak D and Stojanovic G 2015 Sensing mechanism of RuO_2–SnO_2 thick film pH sensors studied by potentiometric method and electrochemical impedance spectroscopy *J. Electroanal. Chem.* **759** 82–90

[34] da Silva, G M, Lemos S G, Pocrifka L A, Marreto P D, Rosario A V and Pereira E C 2008 Development of low-cost metal oxide pH electrodes based on the polymeric precursor method *Anal. Chim. Acta* **616** 36–41

[35] Marzouk S A M, Ufer S, Buck R P, Johnson T A, Dunlap L A and Cascio W E 1998 Electrodeposited iridium oxide pH electrode for measurement of extracellular myocardial acidosis during acute ischemia *Anal. Chem.* **70** 5054–61

[36] Kwon D-H, Cho B-W, Kim C-S and Sohn B-K 1996 Effects of heat treatment on Ta_2O_5 sensing membrane for low drift and high sensitivity pH-ISFET *Sens. Actuators* B **34** 441–5

[37] Zhao R, Xu M, Wang J and Chen G 2010 A pH sensor based on the TiO_2 nanotube array modified Ti electrode *Electrochim. Acta* **55** 5647–51

[38] Zhang W-D and Xu B 2009 A solid-state pH sensor based on WO_3-modified vertically aligned multiwalled carbon nanotubes *Electrochem. Commun.* **11** 1038–41

[39] Yamamoto K, Shi G, Zhou T, Xu F, Zhu M, Liu M, Kato T, Jin J-Y and Jin L 2003 Solid-state pH ultramicrosensor based on a tungstic oxide film fabricated on a tungsten nano-electrode and its application to the study of endothelial cells *Anal. Chim. Acta* **480** 109–17

[40] Razmi H, Heidari H and Habibi E 2008 pH-sensing properties of PbO_2 thin film electro-deposited on carbon ceramic electrode *J. Solid State Electrochem.* **12** 1579–87

[41] Teixeira M F S, Ramos L A, Fatibello-Filho O and Cavalheiro É T G 2001 PbO_2-based graphite–epoxy electrode for potentiometric determination of acids and bases in aqueous and aqueous–ethanolic media *Fresenius' J. Anal. Chem.* **370** 383–6

[42] Telli L, Brahimi B and Hammouche A 2000 , Study of a pH sensor with MnO_2 and montmorillonite-based solid-state internal reference *Solid State Ionics* **128** 255–9

[43] Qingwen L, Guoan L and Youqin S 2000 Response of nanosized cobalt oxide electrodes as pH sensors *Anal. Chim. Acta* **409** 137–42

[44] Mohammad-Rezaei R, Soroodian S and Esmaeili G 2019 Manganese oxide nanoparticles electrodeposited on graphenized pencil lead electrode as a sensitive miniaturized pH sensor *J. Mater. Sci., Mater. Electron.* **30** 1998–2005

[45] Tsai C N, Chou J C, Sun T P and Hsiung S K 2006 Study on the time-dependent slow response of the tin oxide pH electrode *IEEE Sens. J.* **6** 1243–9

[46] Tsai C-N, Chou J-C, Sun T-P and Hsiung S-K 2005 Study on the sensing characteristics and hysteresis effect of the tin oxide pH electrode *Sens. Actuators* B **108** 877–82

[47] Prats-Alfonso E, Abad L, Casañ-Pastor N, Gonzalo-Ruiz J and Baldrich E 2013 Iridium oxide pH sensor for biomedical applications. Case urea–urease in real urine samples *Biosens. Bioelectron.* **39** 163–9

[48] Chiang J-L, Jan S-S, Chou J-C and Chen Y-C 2001 Study on the temperature effect, hysteresis and drift of pH-ISFET devices based on amorphous tungsten oxide *Sens. Actuators* B **76** 624–8

[49] Chen M, Jin Y, Qu X, Jin Q and Zhao J 2014 Electrochemical impedance spectroscopy study of Ta_2O_5-based EIOS pH sensors in acid environment *Sens. Actuators* B **192** 399–405

[50] Schöning M J, Brinkmann D, Rolka D, Demuth C and Poghossian A 2005 CIP (cleaning-in-place) suitable 'non-glass' pH sensor based on a Ta_2O_5-gate EIS structure *Sens. Actuators* B **111–112** 423–9

[51] Chou J C and Wang Y F 2002 Preparation and study on the drift and hysteresis properties of the tin oxide gate ISFET by the sol–gel method *Sens. Actuators* B **86** 58–62

[52] Buck R P and Lindner E 1994 Recommendations for nomenclature of ionselective electrodes (IUPAC Recommendations 1994) *Pure Appl. Chem.* **66** 2527–36

[53] Uppuluri K, Szwagierczak D, Fernandes L, Zaraska K, Lange I, Synkiewicz-Musialska B and Manjakkal L 2023 A high-performance pH-sensitive electrode integrated with a multi-sensing probe for online water quality monitoring *J. Mater. Chem.* C **11** 15512–20

[54] Bousse L and Bergveld P 1984 The role of buried OH sites in the response mechanism of inorganic-gate pH-sensitive ISFETs *Sens. Actuators* **6** 65–78

[55] Liao Y-H and Chou J-C 2009 Preparation and characterization of the titanium dioxide thin films used for pH electrode and procaine drug sensor by sol–gel method *Mater. Chem. Phys.* **114** 542–8

[56] Chou J and Chen C 2009 Fabrication and application of ruthenium-doped titanium dioxide films as electrode material for ion-sensitive extended-gate FETs *IEEE Sens. J.* **9** 277–84

[57] Emami H, Mahinnezhad S, Shboul A A, Ketabi M, Shih A and Izquierdo R 2021 Flexible chemiresistive pH sensor based on polyaniline/carbon nanotube nanocomposite for IoT applications *2021 IEEE Sensors* pp 1–4

[58] Angizi S, Yu E Y C, Dalmieda J, Saha D, Selvaganapathy P R and Kruse P 2021 Defect engineering of graphene to modulate pH response of graphene devices *Langmuir* **37** 12163–78

[59] Lee W S, Park Y-S and Cho Y-K 2014 Hierarchically structured suspended TiO_2 nanofibers for use in UV and pH sensor devices *ACS Appl. Mater. Interfaces* **6** 12189–95

[60] Gou P, Kraut N D, Feigel I M, Bai H, Morgan G J, Chen Y, Tang Y, Bocan K, Stachel J and Berger L 2014 Carbon nanotube chemiresistor for wireless pH sensing *Sci. Rep.* **4** 4468

[61] Lvovich V F 2012 *Impedance Spectroscopy: Applications to Electrochemical and Dielectric Phenomena,* (New York: Wiley)

[62] Manjakkal L, Djurdjic E, Cvejin K, Kulawik J, Zaraska K and Szwagierczak D 2015 Electrochemical impedance spectroscopic analysis of RuO_2-based thick film pH sensors *Electrochim. Acta* **168** 246–55

[63] Manjakkal L, Sakthivel B, Gopalakrishnan N and Dahiya R 2018 Printed flexible electrochemical pH sensors based on CuO nanorods *Sensors Actuators* B **263** 50–8

[64] Manjakkal L, Cvejin K, Kulawik J, Zaraska K, Socha R P and Szwagierczak D 2016 X-ray photoelectron spectroscopic and electrochemical impedance spectroscopic analysis of RuO_2–Ta_2O_5 thick film pH sensors *Anal. Chim. Acta* **931** 47–56

[65] Pejcic B and Marco R D 2006 Impedance spectroscopy: over 35 years of electrochemical sensor optimization *Electrochim. Acta* **51** 6217–29

[66] Lasia A 2002 Electrochemical impedance spectroscopy and its applications *Modern Aspects of Electrochemistry* (Berlin: Springer) pp 143–248

[67] Gao W and Song J 2009 Polyaniline film based amperometric pH sensor using a novel electrochemical measurement system *Electroanalysis* **21** 973–8

[68] Pizzariello A, Stredanský M, Stredanská S and Miertuš S 2001 Urea biosensor based on amperometric pH-sensing with hematein as a pH-sensitive redox mediator *Talanta* **54** 763–72

[69] Neupane S, Subedi V, Thapa K K, Yadav R J, Nakarmi K B, Gupta D K and Yadav A P 2022 An alternative pH sensor: graphene oxide-based electrochemical sensor *Emerg. Mater.* **5** 509–17

[70] Pijanowska D G and Torbicz W 1997 pH-ISFET based urea biosensor *Sens. Actuators* B **44** 370–6

[71] Bergveld P 2003 Thirty years of ISFETOLOGY: what happened in the past 30 years and what may happen in the next 30 years *Sens. Actuators* B **88** 1–20

[72] Chin Y-L, Chou J-C, Sun T-P, Liao H-K, Chung W-Y and Hsiung S-K 2001 A novel SnO_2/Al discrete gate ISFET pH sensor with CMOS standard process *Sens. Actuators* B **75** 36–42

[73] Liao H-K, Chi L-L, Chou J-C, Chung W-Y, Sun T-P and Hsiung S-K 1999 Study on pH pzc and surface potential of tin oxide gate ISFET *Mater. Chem. Phys.* **59** 6–11

[74] Nakazato K 2009 An integrated ISFET sensor array *Sensors* **9** 8831–51

[75] Lee C-S, Kim S K and Kim M 2009 Ion-sensitive field-effect transistor for biological sensing *Sensors* **9** 7111–31

[76] Sinha S and Pal T 2022 A comprehensive review of FET-based pH sensors: materials, fabrication technologies, and modeling *Electrochem. Sci. Adv.* **2** e2100147

[77] Voorthuyzen J and Bergveld P 1990 Photoelectric effects in $Ta_2O_5SiO_2$ Si structures *Sens. Actuators* B **1** 350–3

[78] van der spiegel, J, Lauks I, Chan P and Babic D 1983 The extended gate chemically sensitive field effect transistor as multi-species microprobe *Sens. Actuators* **4** 291–8

[79] Guerra E M, Silva G R and Mulato M 2009 Extended gate field effect transistor using V_2O_5 xerogel sensing membrane by sol–gel method *Solid State Sci.* **11** 456–60

[80] Chen S, Chang S and Chang S 2014 Investigation of InN nanorod-based EGFET pH sensors fabricated on quartz substrate *Digest J. Nanomater. Biostruct.* **9** 1505–11

[81] Campos R D C, Cestarolli D T, Mulato M and Guerra E M 2015 Comparative sensibility study of WO_3 pH sensor using EGFET and cyclic voltammetry *Mater. Res.* **18** 15–9

[82] Hung S-C, Cheng N-J, Yang C-F and Lo Y-P 2014 Investigation of extended-gate field-effect transistor pH sensors based on different-temperature-annealed bi-layer MWCNTs-In$_2$O$_3$ films *Nanoscale Res. Lett.* **9** 1–8

[83] Li H-H, Dai W-S, Chou J-C and Cheng H-C 2012 An extended-gate field-effect transistor with low-temperature hydrothermally synthesized nanorods as pH sensor *IEEE Electron Device Lett.* **33** 1495–7

[84] Cao Q and Rogers J A 2009 Ultrathin films of single-walled carbon nanotubes for electronics and sensors: a review of fundamental and applied aspects *Adv. Mater.* **21** 29–53

[85] Wang K, Liu X, Zhao Z, Li L, Tong J, Shang Q, Liu Y and Zhang Z 2023 Carbon nanotube field-effect transistor based pH sensors *Carbon* **205** 540–5

[86] Pyo J-Y and Cho W-J 2017 High-performance SEGISFET pH sensor using the structure of double-gate a-IGZO TFTs with engineered gate oxides *Semicond. Sci. Technol.* **32** 035015

[87] Loi A, Manunza I and Bonfiglio A 2005 Flexible, organic, ion-sensitive field-effect transistor *Appl. Phys. Lett.* **86** 103512

[88] Li C A, Han K N, Pham X-H and Seong G H 2014 A single-walled carbon nanotube thin film-based pH-sensing microfluidic chip *Analyst* **139** 2011–5

[89] Bartic C, Campitelli A, Baert K, Suls J and Borghs S 2000 Organic-based transducer for low-cost charge detection in aqueous media *Int. Electron Devices Meeting 2000. Technical Digest. IEDM (Cat. No. 00CH37138)* pp 411–4

[90] Lin P and Yan F 2012 Organic thin-film transistors for chemical and biological sensing *Adv. Mater.* **24** 34–51

IOP Publishing

Advanced Electrochemical pH Sensing Technologies
Scientific fundamentals and applications
Libu Manjakkal

Chapter 3

Reference electrodes for electrochemical pH sensors

3.1 Introduction

There is a tremendous increase in the development of new electrochemical devices, including sensors and energy storage. For the fabrication of these devices and for the characterisation of new active electrode materials, one of the essential electrodes is a suitable reference electrode (RE). The RE is one of the most important electrodes in nearly all electrochemical techniques, such as cyclic voltammetry (CV), potentiometry, and electrochemical impedance spectroscopy (EIS) analysis [1, 2]. The RE is essential for developing various electrochemical devices, including sensors for monitoring chemical and biological parameters and conducting electrochemical analyses of the working electrode materials used in different devices, such as sensors, supercapacitors, batteries, and dye-sensitised solar cells [2–7]. In a potentiometric sensor, it acts as a counter electrode to complete the circuit [6, 8, 9]. During an electrochemical reaction, there is no current flow through the RE, which is expected to provide a stable potential without any drift or hysteresis, and with minimal temperature dependence [6, 8, 10]. The major challenges encountered during the fabrication of the RE include achieving compatible dimensions for various applications, ensuring repeatability, and maintaining long-term stability. Different types of REs, their applications, the implementations in various analysis methods and sensors development are summarised in figure 3.1.

It was found that for electrochemical applications, standard hydrogen electrode (SHE), saturated calomel electrodes ($Hg|Hg_2Cl_2$) [11], $Cu|CuSO_4$ [12, 13] and $Ag|AgCl$ [6, 14–16] are the major types of RE used for fabrication or materials performance analysis. Considering the ease of fabrication, stable electrode potential in various conditions, and their non-toxicity, the $Ag|AgCl$-based RE has found

doi:10.1088/978-0-7503-6079-1ch3

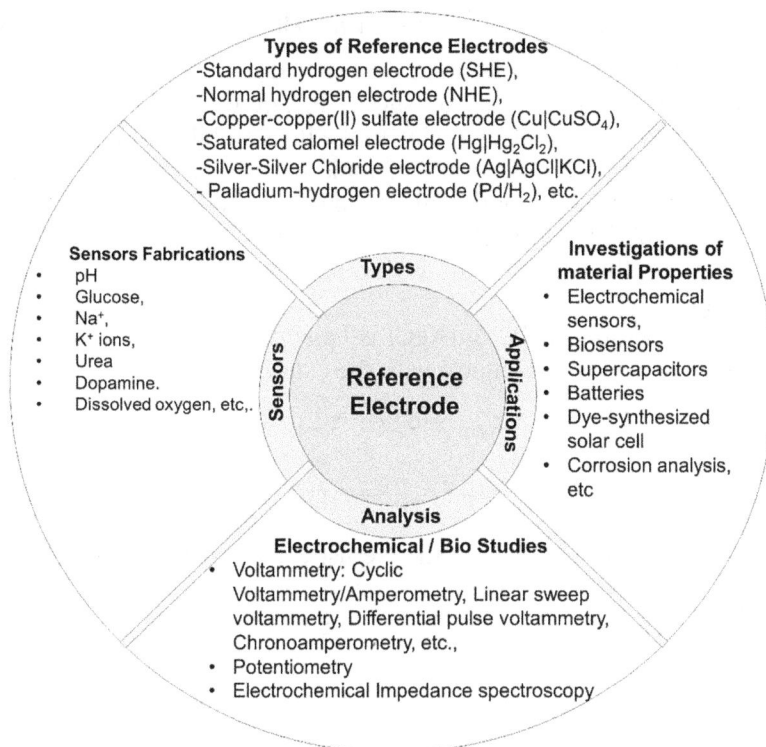

Types of Reference Electrodes
-Standard hydrogen electrode (SHE),
-Normal hydrogen electrode (NHE),
-Copper-copper(II) sulfate electrode ($Cu|CuSO_4$),
-Saturated calomel electrode ($Hg|Hg_2Cl_2$),
-Silver-Silver Chloride electrode ($Ag|AgCl|KCl$),
- Palladium-hydrogen electrode (Pd/H_2), etc.

Types

Sensors Fabrications
- pH
- Glucose,
- Na^+,
- K^+ ions,
- Urea
- Dopamine.
- Dissolved oxygen, etc,.

Sensors

Reference Electrode

Applications

Investigations of material Properties
- Electrochemical sensors,
- Biosensors
- Supercapacitors
- Batteries
- Dye-synthesized solar cell
- Corrosion analysis, etc

Analysis

Electrochemical / Bio Studies
- Voltammetry: Cyclic Voltammetry/Amperometry, Linear sweep voltammetry, Differential pulse voltammetry, Chronoamperometry, etc.,
- Potentiometry
- Electrochemical Impedance spectroscopy

Figure 3.1. Distinct types of REs and their applications in sensor fabrication, investigation of material properties and using different analysis methods. Reproduced from [4]. CC BY 4.0.

excellent applications. The traditional REs are developed based on a rigid glass-based Ag|AgCl electrode configuration, and they show excellent potential stability for long-term applications. Recently, flexible electrochemical sensors or biosensors, including pH, various ions, glucose and lactose monitoring sensors, have attracted significant interest in wearable applications [17, 18]. In addition to this, the deployment of sensors for long-term monitoring, including food or water and soil quality analysis, is required to be compatible with the RE [19–22]. It was found that due to a lack of flexibility, integration difficulties for long-term and multi-sensor, the traditional glass-based REs are limited in many applications. Hence, due to the importance of the development of miniaturised flexible sensors for wearables and for long-term and disposable electrochemical sensors/biosensors, and electrochemical analysis of materials, the recent fabrication of miniaturised REs has attracted significant attention [4, 16, 23–26].

This chapter provides an overview of the design of the glass-based Ag|AgCl RE and its mechanism, the development status of new thick- and thin-film-based REs for various applications. The significance of REs in electrochemical sensors or biosensors analysis is also discussed in this chapter.

3.2 Glass-based Ag|AgCl|KCl REs and their working mechanism

Glass-based Ag|AgCl consists of an Ag wire coated with AgCl and immersed in a 3.5 M or saturated KCl solution, as shown in figure 3.2. The KCl solution is generally considered as an electrolyte due to the high mobility of both K^+ and Cl^-, and it ensures a low impedance path of ionic current between the internal and external test solution [6]. During the fabrication, the ratio between Ag and AgCl is considered as 4:1 [6]. The electrolyte is kept in a refillable glass container with a porous membrane at the tip, which allows ion exchange at the liquid–liquid junction, as shown in figure 3.2.

The electrode reaction of the Ag|AgCl is based on two simultaneous reversible reactions when the electrode contacts with the external solution as given below.

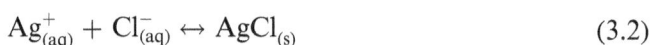

$$Ag^+_{(Aq)} + e^- \leftrightarrow Ag_{(s)} \tag{3.1}$$

$$Ag^+_{(aq)} + Cl^-_{(aq)} \leftrightarrow AgCl_{(s)} \tag{3.2}$$

The electrode potential is defined by the Nernst equation as given below

$$E = E^0 - 2.303\frac{RT}{nF} \log ([Cl^-]) \tag{3.3}$$

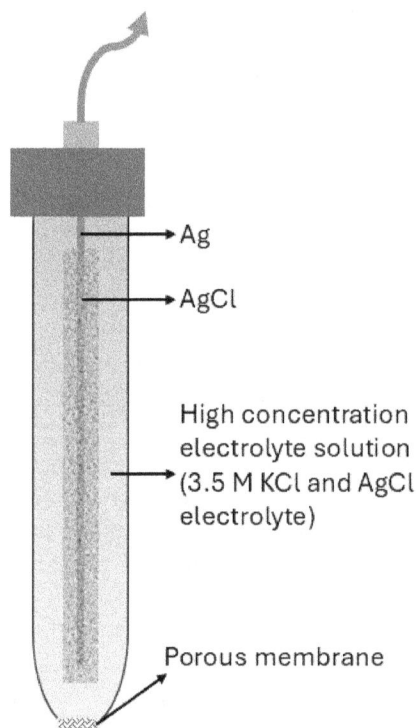

Figure 3.2. Schematic representation of glass-based Ag|AgCl RE.

Here E^0 is the electrode standard potential in V, n is the number of electrons in the electrochemical reaction, R is the molar gas constant, T is the temperature, F is the Faraday constant, and [Cl$^-$] is the concentration of free chloride ions. At the temperature of 25 °C an Ag|AgCl electrode exhibits a susceptibility of −59.16 mV/pCl.

3.3 Thick-film reference electrode based on Ag|AgCl

One of the main disadvantages of advanced functional materials-based planar electrochemical sensors or biosensors, particularly in the context of miniaturised devices for various applications, is the absence of a miniaturised RE. Using glass-based materials in emerging applications such as wearable systems is challenging due to the bulky, rigid, and non-bendable nature of glass. To address these issues, Ag|AgCl-based thin- and thick-film REs have garnered significant interest for wearable and disposable devices [27, 28]. Such planar thick-film REs were also implemented for the biosensing and electrodeposition applications [28]. The electrode geometry also influences the performance of an RE, especially in attaining viable, low-noise, and stable signals [28]. One of the approaches implemented for the design of different geometries of thick-film REs with PDMS + solid NaCl composites as a salt reservoir (which was spin-coated on top of the Ag/AgCl) is shown in figure 3.3 [28].

In RE fabrication, the development Ag electrode layer is crucial for both flexible and non-flexible applications. The potential for miniaturisation through fabrication techniques such as printing technology is promising due to low manufacturing costs, material flexibility, and easy integration with electronic circuits. However, compared

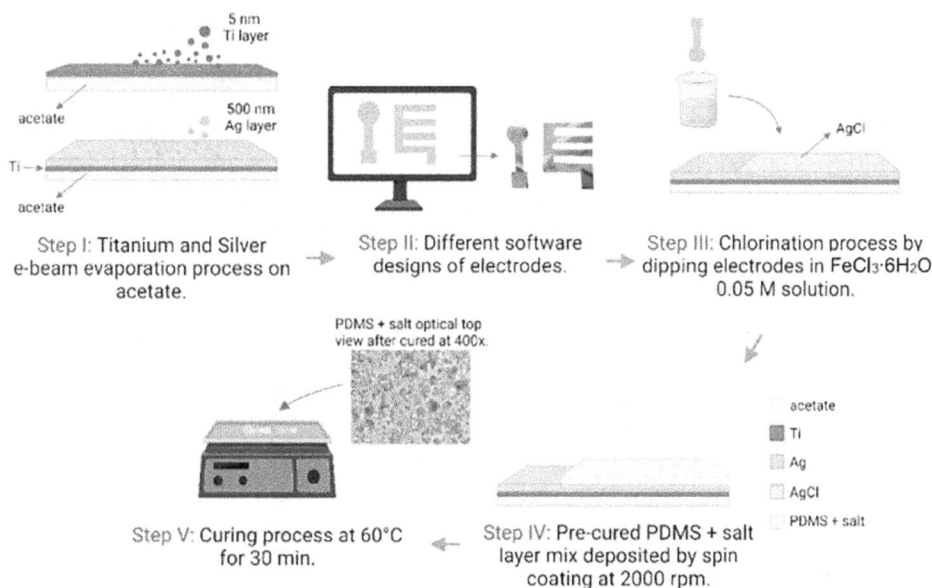

Figure 3.3. Schematic representation of the fabrication of thick-film RE. Reprinted from [28], copyright (2022), with permission from Elsevier.

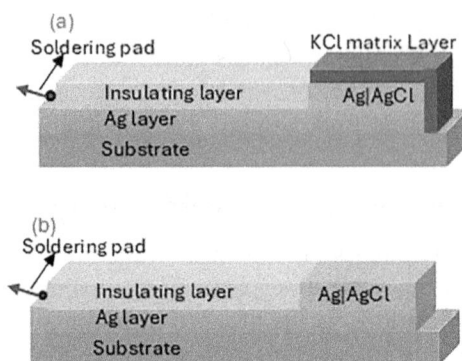

Figure 3.4. (a) RE and (b) quasi-RE.

to full electrodes like Ag|AgCl|KCl, most reported thin- and thick-film REs are quasi-REs (i.e., lacking a KCl layer) [23]. The absence of a salt bridge layer results in fluctuations in the open circuit potential, caused by the gradual decomposition of the AgCl layer over time. Recent research has mainly focused on developing new composite KCl layers printed or coated with Ag|AgCl [6]. A comparison between printed REs and quasi-REs is shown in figure 3.4.

There are various approaches implemented for the development of miniaturised REs, which are provided in detail in one of the review articles by Sophocleous and Atkinson [6]. Among various works, Cranny and Atkinson [29] reported a thick-film RE-based on Ag–AgCl on an alumina substrate. For the electrode fabrication, they used polymer Ag/AgCl, glassy AgCl and glassy Ag/AgCl pastes. The layers deposited using these pastes showed excellent Cl^- ion sensitivity and close to the theoretical Nernstian behaviour. Furthermore, these authors investigated the influence of the hydration port dimensions and the type of sealant used for waterproofing on the RE performance [29]. Tymecki *et al* [30] reported a new approach for the fabrication of inexpensive Ag/AgCl/KCl RE by thick-film technology. The developed electrodes exhibit long operation and storage time and excellent applicability. Sun and Wang [31] reported for the first time a simple procedure for the fabrication of Ag/AgCl electrode on a glass substrate through a combination of electroless deposition and electroplating techniques. The fabricated electrode reveals excellent sensing performance in potentiometric and amperometric applications [31]. It showed a Cl^- ion sensitivity of 54.7 mV/decade (shown in figure 3.5(a)) and a long-term stability. This method of fabrication is inexpensive and can be used for the mass production of disposable devices. Horton *et al* [32] fabricated an Ag|AgCl RE by thick-film technology for a varactor-based inductively coupled wireless pH sensor. For the RE fabrication, these authors converted Ag conducting layer into AgCl by chemical treatment with sodium hypochlorite solution and then overprinted a KCl paste. The fabricated sensor showed good sensing performance for pH measurements [32].

Gao *et al* reported a fully integrated wearable multi-sensor (glucose, lactate, sodium and potassium ions and skin temperature) for sweat analysis [33]. Schematic

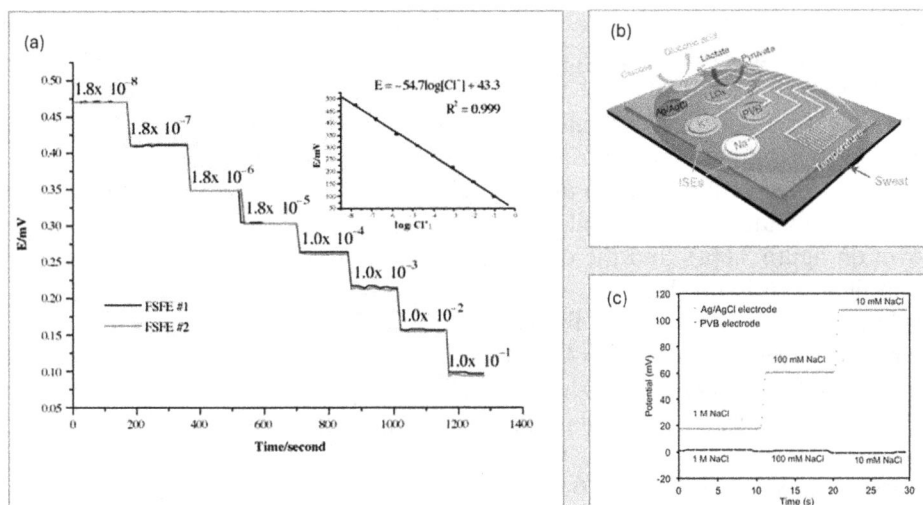

Figure 3.5. (a) Potential response (against a commercial Ag/AgCl RE) of two planar Ag/AgCl REs (working electrodes FSFE #1 and FSFE #2) to different concentrations of Cl⁻. The chloride concentrations (mol l⁻¹) are indicated at each step. Inset: Calibration curve of the average potential response of two planar Ag/AgCl REs to [Cl⁻]. Reprinted from [31], copyright (2006), with permission from Elsevier. (b) A schematic representation of the multi-sensor, which includes solid-state Ag|AgCl RE and a PVB-coated Ag|AgCl RE for multiplexed perspiration analysis. (c) The potential stability of a PVB-coated Ag|AgCl RE and a solid-state Ag|AgCl RE (versus commercial aqueous Ag|AgCl RE) in different NaCl solutions. (b) and (c) reproduced from [33] with permission from Springer Nature.

representation of this multi-sensor is shown in figure 3.5(b). In this multisensory configuration a solid-state Ag|AgCl electrode serves as a shared reference and counter electrode for glucose and lactate sensors. However, for the measurement of Na^+ and K^+ levels, a polyvinyl butyral (PVB)-coated Ag|AgCl RE was employed. In this, PVB was to maintain stable potential in electrolytes with different ionic strengths. In this work, the Ag electrode (prepared by electron-beam evaporation) was chemically treated with $FeCl_3$ solution for chloridizing the electrode [33]. The potential stability of both REs was tested in different NaCl solutions, and the electrode showed excellent stability in different concentrations of solutions, as shown in figure 3.5(c). From figure 3.5(c) it is observed that PVB-based REs have almost stable potential despite changing the concentration of NaCl [33]. The major studies on printed REs show that stable long-term performance, cross-sensitivity, lifetime and potential drift or fluctuations of the RE depends on the type of salt matrix used for the fabrication of the electrode.

3.4 Case study: development of non-flexible and flexible planar Ag|AgCl|KCl reference electrode

Based on the method by Horton *et al* [32], Manjakkal and their team reported fabrication of Ag|AgCl|KCl–glass composite thick-film RE for potentiometric pH sensors on alumina, flexible PET and low temperature co-fired ceramic (LTCC) substrates [4, 8, 9, 34]. For this a new glass–KCl composite paste was developed by using KCl and a lead-free glass powder (containing 87 wt.% Bi_2O_3, 6 wt.% SiO_2, 3 wt.% B_2O_3, 3 wt.% CdO and 1 wt.% Li_2O_3) with a low melting point mixed in equal weight proportions and milled in isopropyl alcohol in a planetary ball mill for 3 h, using agate grinding media. The obtained fine powder was thoroughly mixed in an agate mortar with a solution of ethyl cellulose in terpineol for a rigid planar RE [8]. The schematics of the developed thick-film RE on an alumina substrate are shown on figure 3.6(a), and the device is compared with a commercial glass-based RE in figure 3.6(b) [8]. Initially, an Ag layer was printed on the alumina substrate, dried at 120 °C for 15 min, and fired at 850 °C for 20 min. Then, one end of the Ag layer was converted into an Ag|AgCl layer by reacting with a sodium hypochlorite solution, and the other end of the conducting layer was left for electrical contact. The Ag|AgCl layer, which exhibits strong sensitivity to chloride ion concentration, creates the basic part of the RE. On top of this layer, a glass–KCl composite paste was overprinted and dried at 120 °C for 15 min. Then, the glass–KCl thick film was

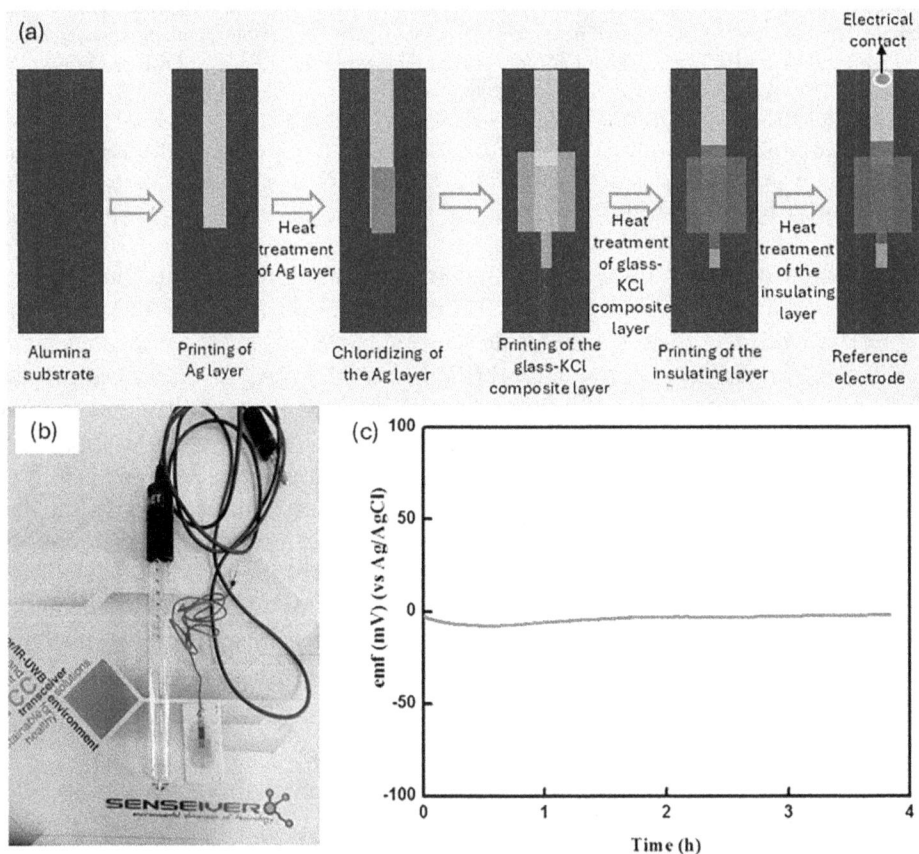

Figure 3.6. (a) A case study for the fabrication of thick-film RE, (b) comparison of commercial glass-based and thick-film REs, and (c) open circuit potential between thick-film and glass-based REs. Reprinted from [8], copyright (2014), with permission from Elsevier.

fired at 550 °C for 1 h [8]. Finally, a protective layer (polyurethane resin) was painted on the top of the substrate surface, leaving uncovered the sensitive electrode (SE) area, pads for electrical contact and a small opening for the hydration port in the KCl layer. This KCl layer functions as an ionic conducting path between the Ag|AgCl and the external solution. The major advantage of the developed KCl–glass composite is that it functions for a relatively longer lifetime as compared to pure KCl powder-based paste. The potential difference between this printed electrode with glass-based RE was measured for both electrodes dipped in deionised water and found to be close to zero, which implies the proper operation of the printed electrode, as given in figure 3.6(c). The shape and composition of the glass–KCl layer and the hydration port may influence the potential and lifetime of the RE. Optimization of these factors is important to restrict the salt loss from the layer [8].

For a flexible pH sensor fabrication, a flexible RE (FRE) was developed by using a similar glass–KCl powder composite with changes in the paste preparation [4]. To prepare the past for FRE, the glass–KCl powder composite from above was mixed with a binder made of copolymer of poly (methyl methacrylate) (PMMA), poly (butyl methacrylate) (PMBA), and butyl carbitol acetate (BCA) as a solvent. The fabrication steps of the FRE and the image of the developed FRE are given in figure 3.7.

The developed FRE shows an open circuit potential of 68 mV against glass glass-based RE (figure 3.8(a)). This FRE application in pH sensing was also evaluated and shows that the FRE has a stable potential value of −31 mV against a pH-SE. The performances were also compared with glass-based RE and pH SE, and they show a potential of 30 mV. Further, the bending appears to have little influence on the

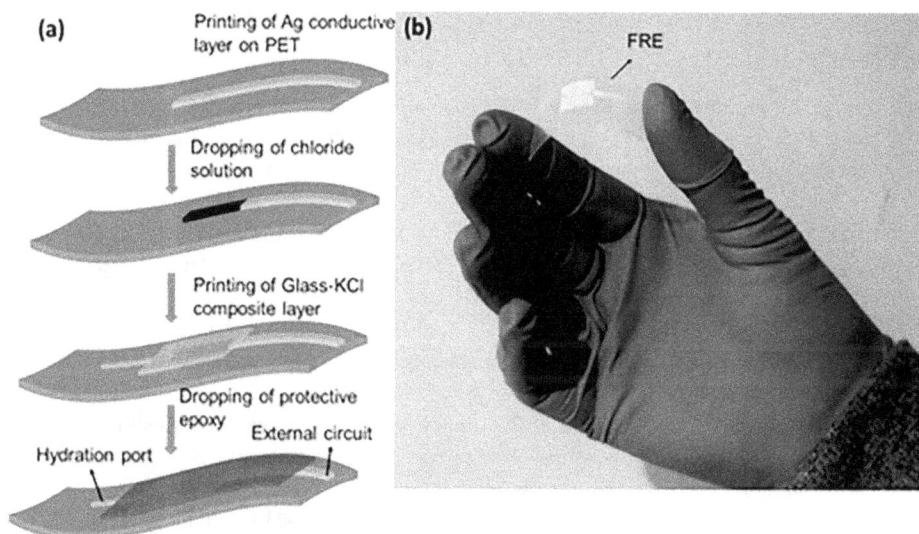

Figure 3.7. Fabrication steps of FRE and (b) image of FRE. Reproduced from [4] CC BY 4.0.

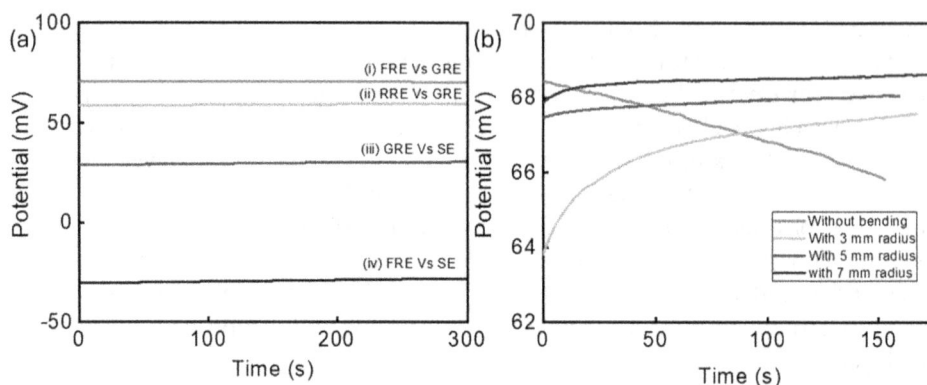

Figure 3.8. The potential between the (i) FRE versus glass RE, (ii) glass RE versus SE, (iii) FRE versus SE, and (b) the potential variation with and without bending of FRE. Reproduced from [4] CC BY 4.0.

potential values, as the variation with bending is low (±4 mV) (figure 3.8(b)). This performance shows excellent application of the thick-film RE in electrochemical and biosensing applications [4].

3.5 Summary

REs carry significant importance in the fabrication of electrochemical and bio-sensors, and electrochemical characterisation of active working electrode materials of energy storage devices such as batteries and supercapacitors. Fabrication of a stable and known potential reference is essential in electrochemistry. It requires high efficiency and long-term stability in different conditions. This chapter discusses the importance of REs, and various configurations and new designs of miniaturised REs.

References

[1] Inzelt G, Lewenstam A and Scholz F 2013 *Handbook of Reference Electrodes* (Springer)
[2] Troudt B K, Rousseau C R, Dong X I, Anderson E L and Bühlmann P 2022 Recent progress in the development of improved reference electrodes for electrochemistry *Anal. Sci.* **38** 71–83
[3] Gromadskyi D G 2016 Characterization of Ag/Ag$_2$SO$_4$ system as reference electrode for in-situ electrochemical studies of advanced aqueous supercapacitors *J. Chem. Sci.* **128** 1011–7
[4] Manjakkal L, Shakthivel D and Dahiya R 2018 Flexible printed reference electrodes for electrochemical applications *Adv. Mater. Technol.* **3** 1800252
[5] Tang Y-b, Wang S-n, Chen L and Fan Z-h 2015 Preparation and properties of embeddable Ag/AgCl gelling reference electrode for rebars corrosion monitoring in concrete *China Ocean Eng.* **29** 925–32
[6] Sophocleous M and Atkinson J K 2017 A review of screen-printed silver/silver chloride (Ag/AgCl) reference electrodes potentially suitable for environmental potentiometric sensors *Sens. Actuators* A **267** 106–20
[7] Zaban A, Zhang J, Diamant Y, Melemed O and Bisquert J 2003 Internal reference electrode in dye sensitized solar cells for three-electrode electrochemical characterizations *J. Phys. Chem.* B **107** 6022–5

[8] Manjakkal L, Cvejin K, Kulawik J, Zaraska K, Szwagierczak D and Socha R P 2014 Fabrication of thick film sensitive RuO_2–TiO_2 and Ag/AgCl/KCl reference electrodes and their application for pH measurements *Sens. Actuators* B **204** 57–67

[9] Manjakkal L, Zaraska K, Cvejin K, Kulawik J and Szwagierczak D 2016 Potentiometric RuO_2–Ta_2O_5 pH sensors fabricated using thick film and LTCC technologies *Talanta* **147** 233–40

[10] Yin J, Qi L and Wang H 2012 Antifreezing Ag/AgCl reference electrodes: fabrication and applications *J. Electroanal. Chem.* **666** 25–31

[11] Yosypchuk B and Novotný L 2004 Reference electrodes based on solid amalgams *Electroanalysis* **16** 238–41

[12] Dewi M S, Alva S and Jamil W A W 2021 Development and characterization of solid Cu/$CuSO_4$ reference electrodes *Int. J. Innov. Mech. Eng. Adv. Mater.* **3** 17–25

[13] Hall D M, Beck J R, Brand E, Ziomek-Moroz M and Lvov S N 2016 Copper–copper sulfate reference electrode for operating in high temperature and high pressure aqueous environments *Electrochim. Acta* **221** 96–106

[14] Rohaizad N, Mayorga-Martinez C C, Novotný F, Webster R D and Pumera M 2019 3D-printed Ag/AgCl pseudo-reference electrodes *Electrochem. Commun.* **103** 104–8

[15] Matsumoto T, Ohashi A and Ito N 2002 Development of a micro-planar Ag/AgCl quasi-reference electrode with long-term stability for an amperometric glucose sensor *Anal. Chim. Acta* **462** 253–9

[16] Cumba L R, Byrne R, Maolmhuaidh F Ó, Morrin A and Forster R J 2024 Greener alternative screen printable ink formulation for the development of flexible reference electrodes *Electrochim. Acta* **501** 144797

[17] Dang W, Manjakkal L, Navaraj W T, Lorenzelli L, Vinciguerra V and Dahiya R 2018 Stretchable wireless system for sweat pH monitoring *Biosens. Bioelectron.* **107** 192–202

[18] Gao W, Emaminejad S, Nyein H Y Y, Challa S, Chen K, Peck A, Fahad H M, Ota H, Shiraki H and Kiriya D 2016 Fully integrated wearable sensor arrays for multiplexed *in situ* perspiration analysis *Nature* **529** 509

[19] Mamilla S, Yousuf S Z, Avulapati M M, Mallahi A A and Murty N V L N 2024 Planar Ag/AgCl reference electrode for electrochemical agronomic sensors *IEEE Sens. Lett.* **8** 1–4

[20] Kageyama T, Hara S, Sarno R, Matsunaga T and Lee S S 2024 Novel sensor using ISFET and Pt electrodes for water pH and flow speed measurement *IEEE Sens. Lett.* **8** 1–4

[21] Uppuluri K, Szwagierczak D, Fernandes L, Zaraska K, Lange I, Synkiewicz-Musialska B and Manjakkal L 2023 A high-performance pH-sensitive electrode integrated with a multi-sensing probe for online water quality monitoring *J. Mater. Chem.* C **11** 15512–20

[22] Markapudi P R, Beg M, Kadara R O, Paul F, Kerrouche A, See C H and Manjakkal L 2025 Nitrate pollution mapping for reservoirs using flexible sensors integrated with underwater robot *IEEE Internet Things J.* **12** 39172–80

[23] Sun J, Mei Y, Bai W, Han L, Li Y, Gao Y, Lang M-F and Xue H 2025 Flexible Ag/AgCl quasi-reference electrode with nano silver dendrites semi-embedded in PDMS: ensuring long operation lifespan, remarkable stability for electrophysiological monitoring and flexible sensor applications *Chem. Eng. J.* **507** 160790

[24] Mandjoukov B and Lindfors T 2024 Planar, low-cost, flexible, and fully laminated graphene paper pseudo-reference and potassium-selective electrodes *Sens. Actuators* B **403** 135190

[25] Wang T, Yao S, Shao L-H and Zhu Y 2024 Stretchable Ag/AgCl nanowire dry electrodes for high-quality multimodal bioelectronic sensing *Sensors* **24** 6670

[26] Moreirinha C, Wittendorp P, Dahl-Hansen R, Jain S, Mielnik M M, Gomes M T S, Rudnitskaya A and Skottvoll F S 2024 Inkjet-printed planar silver/silver chloride pseudo reference electrode for microfluidic potentiometric sensor applications *IEEE Sens. J.* **24** 40351–7

[27] Zhao Z, Tu H, Kim E G, Sloane B F and Xu Y 2017 A flexible Ag/AgCl micro reference electrode based on a parylene tube structure *Sens. Actuators* B **247** 92–7

[28] Torres-González V, Ávila-Niño J A and Araujo E 2022 Facile fabrication of tailorable Ag/AgCl reference electrodes for planar devices *Thin Solid Films* **757** 139413

[29] Cranny A and Atkinson J K 1998 Thick film silver-silver chloride reference electrodes *Meas. Sci. Technol.* **9** 1557

[30] Tymecki Ł, Zwierkowska E and Koncki R 2004 Screen-printed reference electrodes for potentiometric measurements *Anal. Chim. Acta* **526** 3–11

[31] Sun X and Wang M 2006 Fabrication and characterization of planar reference electrode for on-chip electroanalysis *Electrochim. Acta* **52** 427–33

[32] Horton B E, Schweitzer S, DeRouin A J and Ong K G 2011 A varactor-based, inductively coupled wireless pH sensor *IEEE Sens. J.* **11** 1061–6

[33] Gao W *et al* 2016 Fully integrated wearable sensor arrays for multiplexed *in situ* perspiration analysis *Nature* **529** 509–14

[34] Manjakkal L, Cvejin K, Kulawik J, Zaraska K, Szwagierczak D and Stojanovic G 2015 Sensing mechanism of RuO_2–SnO_2 thick film pH sensors studied by potentiometric method and electrochemical impedance spectroscopy *J. Electroanal. Chem.* **759** 82–90

IOP Publishing

Advanced Electrochemical pH Sensing Technologies
Scientific fundamentals and applications
Libu Manjakkal

Chapter 4

Metal oxide-based pH sensors: sensor fabrication and mechanism

4.1 Introduction

The realisation of high-performance electrochemical devices, such as for energy storage and sensors are some of the greatest scientific challenges due to their high demand for functional materials. Metal oxide (MOx) nanostructures are found to be promising active electrode materials for these devices [1–4]. The MOx nanostructures offer a high surface-to-volume ratio, tuneable surface morphology, high electrochemically active surface area, and high stability in performance. The micro- and nanostructured morphologies of MOx lead to its applications in electrochemical-biomedical sciences and across numerous other scientific disciplines. The electrochemical, biocompatible and electrical properties of the nanostructured MOx make them highly suitable for electrochemical sensing. Moreover, MOx enhances the sensitivity, selectivity, and catalytic properties of electrochemical and biosensors [3, 5]. Due to these unique properties, these materials make them highly suitable for the fabrication of sensors for various applications, including water and food quality monitoring, wearable systems for chronic diseases, and industrial applications [3, 6–8]. Among various sensors, the miniaturised electrochemical pH sensors have seen an increasing demand for the next generation of reliable sensors for online monitoring of a solution. In the past few decades, there has been more concentration on measuring the pH of low- and high-temperature aqueous solutions by using different MOx materials. Semiconductor sensing MOx materials help fabricate a pH sensor for online monitoring of the pH of a solution [9].

The main advantages of MOx-based pH sensors are: (i) low cost and simple fabrication; (ii) high sensitivity approaching the Nernstian response over a wide pH range; (iii) rapid response; (iv) extended lifespan; (v) good reproducibility; (vi) low hysteresis and drift effects; (vii) minimal cross-sensitivity to interfering ions such as Li^+, Na^+, and K^+; (viii) potential for large-scale manufacturing; and (ix) suitability

doi:10.1088/978-0-7503-6079-1ch4
4-1

for use in hazardous, high-temperature, and high-pressure environments [9–12]. Various factors, including fabrication methods and material properties, influence the sensing performance of electrochemical MOx-based pH sensors. Researchers have focused on utilising different types of MOx as sensing materials and exploring various preparation and measurement techniques for their development. Notably, both thin and thick-film pH sensors based on different MOx have garnered significant interest. In MOx-based pH sensors, key techniques such as potentiometry, cyclic voltammetry (CV), and electrochemical impedance spectroscopy (EIS) are employed to gain vital insights into the activity mechanisms of sensitive materials. A summary of the materials, fabrication methods, characterisation, and applications are provided in figure 4.1.

The importance of MOx as materials for pH sensors and their sensing mechanisms was first highlighted by Fog and Buck [13]. Until now, different forms of MOx have been used for the construction of a pH sensor. The MOx investigated for pH sensor applications is RuO_2, IrO_2, Ta_2O_5, TiO_2, SnO_2, CeO_2, OsO_2, Co_2O_3, TiO–SnO–SnO_2, Bi_2O_3–Nb_2O_5, WO_3, PbO_2, IrO_2–TiO_2, RuO_2–TiO_2, perovskite lithium

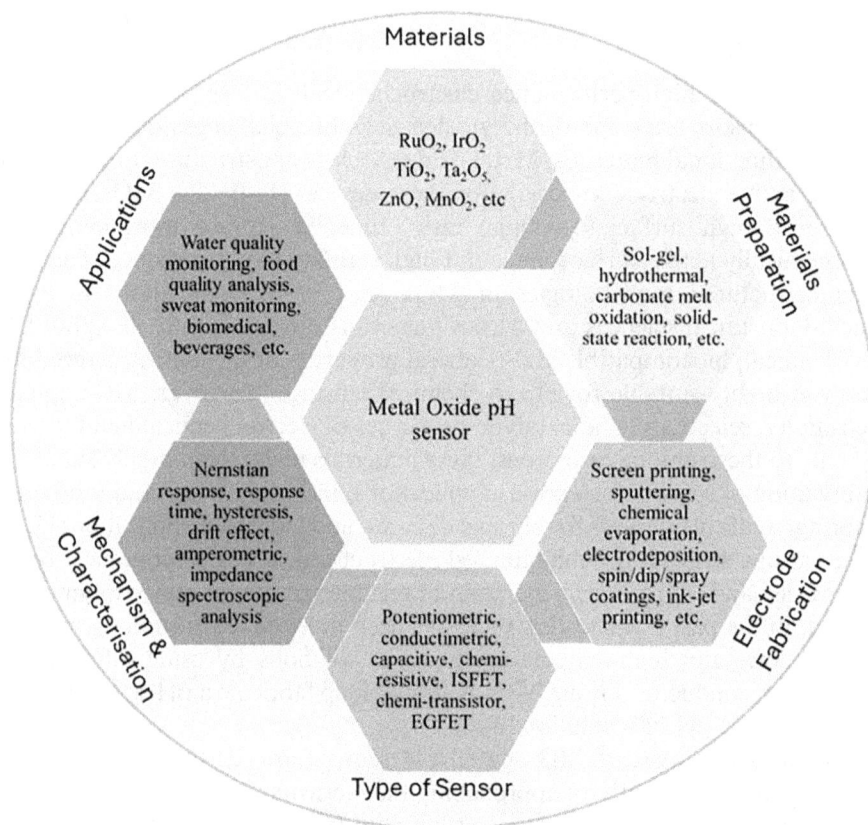

Figure 4.1. Various metal oxide-based pH sensors—key methods for preparation of electrode materials and sensor fabrication, type of sensors, parameters for analysis and applications. Reproduced from [9] CC BY 4.0.

lanthanum titanate (LLTO), yitria stabilized zirconia (YSZ), molybdenum oxide bronzes, etc [9, 10, 12, 14, 15]. Moreover, much work has also been done in metal/MOx-based pH sensors [14]. Some solid-state materials, like those based on antimony and bismuth, exhibit a high sensitivity for hydrogen ions [12, 16, 17]. The redox equilibrium between the metal–MOx phases (e.g. Sb–Sb_2O_3) is the source of their pH sensitivity [12, 16, 17]. However, the performance of such systems in pH measuring is limited, because they are sensitive to several redox agents, as well as showing a systematic deviation from the expected behaviour. Horton *et al* [17] reported Sb/Sb_2O_3-based inductively coupled wireless pH sensor for remote monitoring of pH during industrial or food processes. However, one of the major challenges relevant to MOx electrochemical pH sensors is the explanation of their sensing mechanism, this issue remaining ambiguous.

There is a strong demand for the development and investigation of new low-cost sensitive materials, new sensor designs and new measurement methods for the next generation of miniaturised pH sensors. Moreover, further efforts are necessary to elucidate the pH-sensing mechanism. To meet these challenges, this chapter discusses the pH-sensing mechanism of MOx and the major fabrication approaches.

4.2 Sensing mechanism of electrochemical solid-state materials-based sensors

Over the past decade, although researchers have put considerable effort into understanding the sensing mechanism of metal oxide-based pH sensors, it remains unclear. The growing demand for pH sensing has driven the development of new, stable pH sensors suitable for various conditions. However, designing and developing such sensors requires a clear understanding of the primary mechanisms involved.

Fog and Buck [13] considered five different mechanisms to explain the pH sensitivity of a metal oxide:

- Simple ion exchange on a surface layer containing –OH groups.
- A redox equilibrium between two different solid phases.
- A redox equilibrium involving only one solid phase, whose hydrogen content can be continuously varied by passing current through the electrode.
- A single-phase 'oxygen intercalation' of the electrode.
- Steady state corrosion of the electrode material.

Among these options, Fog and Buck [13] concluded that 'oxygen intercalation' is the most probable mechanism for pH sensitivity. McMurrray *et al* [18] also indicated such a mechanism. The general sensing surface mechanism can be expressed as:

$$MO_x + 2\delta H^+ + 2\delta e^- \Leftrightarrow MO_{x-\delta} + \delta H_2O \tag{4.1}$$

where MO_x is a higher metal oxide and $MO_{x-\delta}$ is a lower metal oxide. The electrode potential E is given by:

$$E = \frac{RT}{F} \ln a_{H^+}^1 + \frac{RT}{2F} \ln a_0^s + \text{constant} \tag{4.2}$$

where $a_{H^+}^l$ is the proton activity in liquid phase, a_0^s is the oxygen activity in solid phase, T is absolute temperature, R is universal gas constant and F is the Faraday constant.

Furthermore, based on the Fog and Buck [13] explanation, Trasatti [19] and Mihell and Atkinson [20] suggested that the pH response could be due to ion exchange in the surface layer containing –OH groups. The relevant electrode potential is given by:

$$E = \frac{RT}{F} \ln [H^+] + \frac{RT}{\delta F} \ln \left[\frac{MO_x(OH)_y}{MO_{x-\delta}(OH)_{y+\delta}} \right] + \text{constant} \qquad (4.3)$$

In general, when the sensor comes into contact with a solution, due to the dissociative adsorption of water, the metal oxide surface is covered by hydroxide groups, which attempt to form oxide sites by releasing protons. This process can result in the formation of a couple of higher and lower valency metal oxides. It leads to the generation of a potential difference between the reference and sensing electrodes. The magnitude of this potential is proportional to the pH of a solution according to the Nernst equation [10]:

$$E = E_0 - \frac{2.303RT}{nF}\text{pH} \qquad (4.4)$$

Here E is electromotive force (emf) of the electrochemical cell, E_0 is standard potential, R is universal gas constant, T is temperature, n is the number of electrons and F is Faraday's constant. At temperature T equal to 25 °C the potential is:

$$E = E_0 - \frac{59.14\,\text{mV}}{n}\text{pH} \qquad (4.5)$$

For one electron electrode reaction, the slope factor of the plot given by equation (4.5) is 59.14 mV/pH.

When a metal oxide is exposed to a solution, on the surface of the oxide, many chemical reactions can occur, like hydration, dissolution, hydrolysis and dissociation [9, 10].

According to the site-binding theory, for the majority of oxides, after immersion in an aqueous solution, the surface groups –O⁻, –OH and –OH₂⁺ are developed [21–23], as illustrated in figure 4.2(a). Protons and hydroxide ions from the solution are

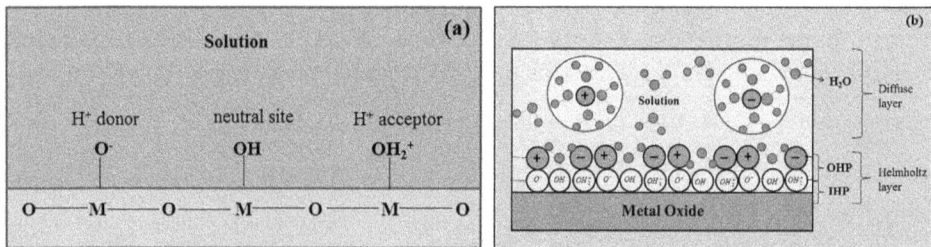

Figure 4.2. (a) Schematic representation of site-binding theory for the metal oxide surface-solution system and (b) electrical double layer at the metal oxide–solution interface.

attracted to oxygen ions in the metal oxide crystal lattice and to the surface cations, respectively. This results in the covering of the MOx by hydroxide groups. The created metal hydroxide groups can donate a proton to the solution and form a negative surface group or accept a proton from the solution, converting into a positive surface group [22, 24, 25].

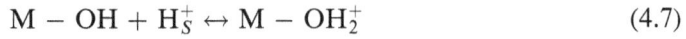

$$M - O^- + H_S^+ \leftrightarrow M - OH \tag{4.6}$$

$$M - OH + H_S^+ \leftrightarrow M - OH_2^+ \tag{4.7}$$

The dissociation constants k_a and k_b can be expressed as:

$$k_a = \frac{[M - O^-][H^+]_S}{[M - OH]} \tag{4.8}$$

$$k_b = \frac{[M - OH][H^+]_S}{[M - OH_2^+]} \tag{4.9}$$

Here $[M-O^-]$, $[M-OH]$ and $[M-OH_2^+]$ groups represent the negative surface groups, neutral sites and the positive surface groups, respectively. H^+ is a proton at the metal oxide–electrolyte interface. Hence, due to these protonation and deprotonation reactions of the surface groups, H^+ and OH^- ions are the potential-determining ions with Nernstian dependence on the solution pH.

The charged surface groups at the metal oxide–solution interface create an electrical double layer (edl) structure [22–25] shown in figure 4.2(b). The edl consists of the Helmholtz (compact layer) and the diffuse layer (Stern layer). The Helmholtz layer contains the inner Helmholtz plane (IHP) and the outer Helmholtz plane (OHP), as illustrated in figure 4.2(b). The charged surface groups are located in IHP and strongly adhere to the metal oxide surface by chemical bonding. In OHP, the ions from the solution are attracted to the surface charge by Coulombic interaction. In the diffuse layer, ions in the solution move freely due to thermal or electrical forces [22–27].

A change in pH value of the solution affects the equilibrium state of the MOx surface and leads to a change in the electrical properties of the sensitive electrode. These reactions may occur due to the involvement of adsorption of ions or diffusion of ions or both adsorption of ions at the MOx surface and their diffusion through the interface into the oxide layer [28]. The variation of electrical properties of edl is dependent on the nature of the electrode material and the chemical reactivity at the electrode–solution interface.

When a sensitive electrode is in contact with the solution, the pH value of the solution affects the equilibrium state of MO_x surface. This leads to changes in electrical properties and surface potential (Ψ) of the electrode [10, 29–32]. The variation in electrical properties of edl depends on the nature of the electrode material and its chemical reactivity at the solution interface. To investigate the sensing mechanism of an ion-sensitive field-effect transistor, Liao et al [31] carried out a detailed experimental and theoretical study based on site-binding theory at the point of zero charge pH_{PZC} and surface potential of a tin oxide gate ISFET. The changes of surface potential (Ψ) of MO_x can be described as [31]:

$$2.303(pH_{PZC} - pH) = \frac{q\psi}{kT} + \sinh^{-1}\left[\frac{q\psi}{kT}\frac{1}{\beta}\right] \qquad (4.10)$$

where, β reflects the sensitivity of the gate insulators and depends on the surface density of hydroxyl groups.

$$\beta = \frac{2q^2 N_s (K_a K_b)^{1/2}}{kT C_{DL}} \qquad (4.11)$$

Here, K_a and K_b are dissociation constants related to acidic and basic equilibrium, N_s is the total number of surface sites per unit area. The pH sensitivity is higher, and the response is more linear when N_s is larger. The value of N_s depends on the crystalline structure of the material. C_{DL} is a simple capacitance derived from the Gouy–Chapman–Stern model.

The surface properties and morphologies of MOx also influence the sensing performance. It was observed the nanostructures of ZnO, such as nanorods, nanowires and nanotubes, have different influences on the pH-sensing performance [33]. Fulati et al [34] developed a new MOx-based pH sensor using ZnO nanotubes and nanorods. The sensor fabricated using ZnO nanotubes had a sensitivity (45.9 mV/pH) as compared to 28.4 mV/pH for the ZnO nanorods-based sensor, in the pH range 4–12. This significant difference in sensitivity occurred because small dimensionally ZnO nanotubes have a higher level of surface and subsurface oxygen vacancies compared to nanorods. ZnO nanotubes exhibit a large effective surface area with a higher surface-to-volume ratio, as compared to ZnO nanorods. The difference in surface charge distribution in two walls of the MOx–electrolyte interface for ZnO nanorods and nanotubes, in the $CaCl_2$ electrolyte, is shown in figures 4.3(a) and (b), respectively [34]. A detailed investigation on the mechanism of pH-sensing using ZnO nanorods, with an emphasis on the nano-interface mechanism, has been described by Hilli and Magnus [22]. A similar observation on morphology influences on sensing performances was found for a CuO-based conductimetric pH sensor. The hydrothermally synthesized CuO formed in different shapes and morphologies, such as CuO nanoflower (CuO–NF) and CuO nanorectangle (CuO–NR), as shown in figures 4.3(c) and (d) [35]. From the EIS analysis of the materials with various pH solutions the CuO–NR-based sensors have lower impedance value with respect to the CuO–NF-based sensor, as shown in figure 4.3(e) [35]. It was observed that the CuO–NR-based sensor exhibits a sensitivity of 0.64 µF/pH in the range of pH 5–8.5, as shown in figure 4.3(f).

4.3 Influence of fabrication on pH-sensing performance

So far, different methods of fabrication have been investigated for the development of an effective metal oxide pH sensor. Most of the sensitive electrodes reported have been prepared in the form of bulk (pellets), thick- and thin-film shapes.

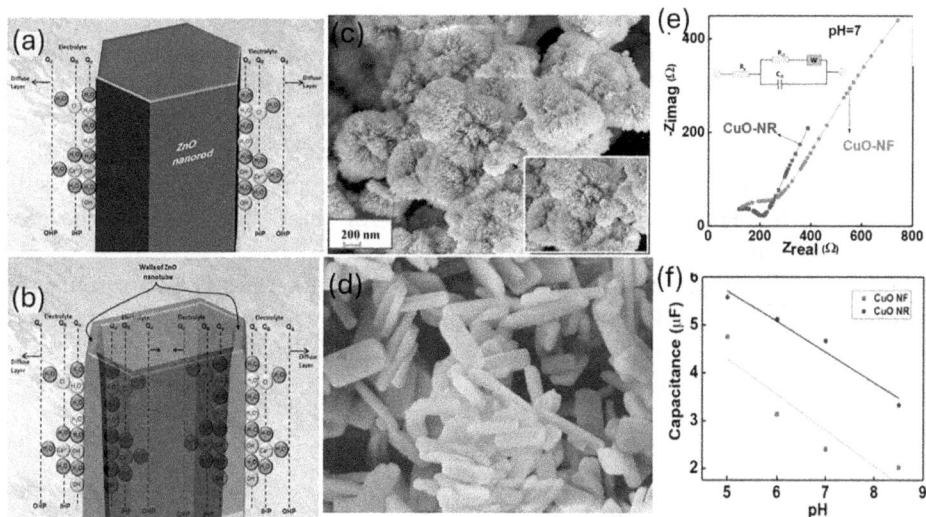

Figure 4.3. (a) and (b) Schematic representation of the ZnO-nanorod and ZnO nanotube electrolyte interaction and its surface charge distribution, respectively. Reproduced from [34] CC BY 3.0. (c) and (d) SEM images of CuO nanoflower and nanorectangle. (e) EIS analysis presented in the Nyquist plot for CuO nanorectangle and nanoflower at pH 7 solution, with an inset showing the equivalent circuit. (f) The change in capacitance of the CuO pH sensor in the region of pH 5–8.5. (c)–(f) reprinted from [35], copyright (2018), with permission from Elsevier.

The fabrication methods to be used will influence the sensing performance due to the variation of structure and electronic characteristics of the pH-sensitive electrode. The possible fabrication methods for the metal oxide film for pH sensors are: screen printing, sol–gel, sputtering, electrochemical oxidation, electrodeposition, carbonate melt oxidation, Pechini method.

The majority of the reported articles are based on screen printing, sputtering and sol–gel methods. One of the most distinguishing and low-cost methods widely used for thick-film-based pH sensor fabrication is the screen-printing method. This method helps to deposit a paste with a miniaturised and precise layout for the development of both sensitive and reference electrodes of a pH sensor. Moreover, this method permits the fabrication of both potentiometric and conductometric-based pH sensors. The screen-printing method offers several advantages over other methods, such as relatively low cost, high performance, high reliability, few solder joints, precise layout, different shapes and miniaturised design destined for multi-sensor and large-scale production of sensors [20, 36]. The sputtering method was mostly used to form high-performance metal oxide thin-film pH sensors. The major steps of sputtering methods are: (i) the material to be deposited is converted into vapour; (ii) the vapour from the material is transferred to the substrate through high vacuum; and (iii) the vapour condenses onto the substrate surface forming a thin film. This method enables the thickness of the film by controlling the deposition time [37, 38]. The sol–gel method produced a metal oxide film of uniform thickness with porous structure and may have very high specific surface areas. In this method, the metal oxide sol can be deposited on the substrate electrode by spray coating or by the dip/spin coating method. To get a sensitive ion-selective electrode, the film is processed with suitable thermal treatment [39, 40]. Figure 4.4 shows the various parameters of pH sensors fabricated by different approaches. From the sensing performances and methods, it was found that the screen-printing method offers many advantageous properties as compared to other approaches. This offers better low-cost electrochemical pH-sensing fabrication for the development of the next-generation of sensors [9].

The sensitivity observed for various methods of fabrication is also summarized and compared with the ideal Nernstian response. Figure 4.5 represents the variation in sensitivity of the sensors for different materials fabricated by screen-printing [41–45], sputtering [38, 46–49], sol–gel [21, 39, 50–52] and electrodeposition [53–55]

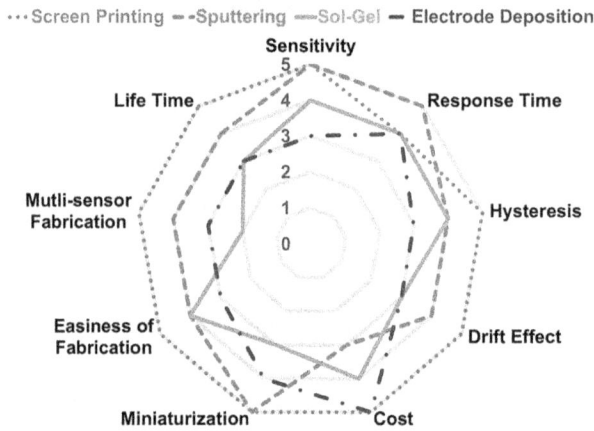

Figure 4.4. Method of fabrication of electrochemical MOx-based pH sensors. Reproduced from [9] CC BY 4.0.

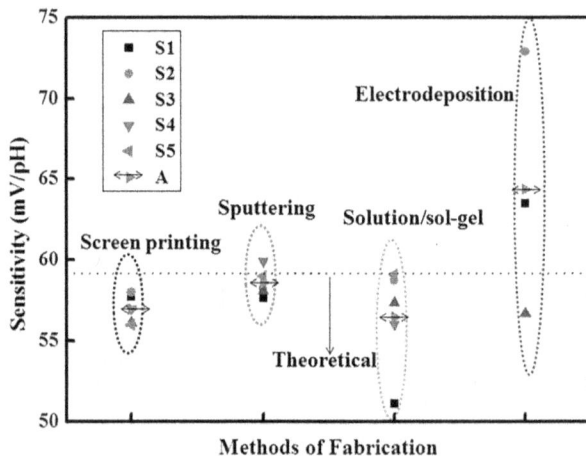

Figure 4.5. Comparison of pH sensitivity of different MOx materials fabricated using different methods and comparison with the theoretical slope of the Nernstian relationship. Reproduced from [9] CC BY 4.0.

methods. These studies reveal that the sensitivity and its mechanism of MO_x pH sensor depend not only on material composition but also on the deposition method. The deposition methods can influence microstructure, porosity, surface homogeneity and crystalline structure of the material.

4.4 Summary

In electrochemical pH sensing, ion-sensitive metal oxides have received substantial attention as sensitive electrodes, attributable to their elevated sensitivity, selectivity, stability, and prompt response. This chapter explores the significance of sensing mechanism and fabrication approaches in MOx-based pH sensors. The chapter also provides information about the influences of surface morphology on the sensing performance of a pH sensor.

Parts of this chapter have been reproduced from [9]. CC BY 4.0.

References

[1] Yan S, Abhilash K P, Tang L, Yang M, Ma Y, Xia Q, Guo Q and Xia H 2019 Research advances of amorphous metal oxides in electrochemical energy storage and conversion *Small* **15** 1804371

[2] Tajik S *et al* 2022 Applications of non-precious transition metal oxide nanoparticles in electrochemistry *Electroanalysis* **34** 1065–91

[3] George J M, Antony A and Mathew B 2018 Metal oxide nanoparticles in electrochemical sensing and biosensing: a review *Microchim. Acta.* **185** 358

[4] Habib W, Saji A, Paul F, Markapudi P R, Wilson C and Manjakkal L 2025 Flexible electrochemical capacitors based on ZnO–carbon black composite *Results Eng.* **25** 104510

[5] Dong Q, Ryu H and Lei Y 2021 Metal oxide based non-enzymatic electrochemical sensors for glucose detection *Electrochim. Acta.* **370** 137744

[6] Rabak A, Uppuluri K, Franco F F, Kumar N, Georgiev V P, Gauchotte-Lindsay C, Smith C, Hogg R A and Manjakkal L 2023 Sensor system for precision agriculture smart watering can *Results Eng.* **19** 101297

[7] Uppuluri K, Szwagierczak D, Fernandes L, Zaraska K, Lange I, Synkiewicz-Musialska B and Manjakkal L 2023 A high-performance pH-sensitive electrode integrated with a multi-sensing probe for online water quality monitoring *J. Mater. Chem.* C **11** 15512–20

[8] Yoon Y, Truong P L, Lee D and Ko S H 2022 Metal-oxide nanomaterials synthesis and applications in flexible and wearable sensors *ACS Nanosci. Au.* **2** 64–92

[9] Manjakkal L, Szwagierczak D and Dahiya R 2020 Metal oxides based electrochemical pH sensors: current progress and future perspectives *Prog. Mater. Sci.* **109** 100635

[10] Kurzweil P 2009 Metal oxides and ion-exchanging surfaces as pH sensors in liquids: State-of-the-art and outlook *Sensors* **9** 4955–85

[11] Vonau W and Guth U 2006 pH monitoring: a review *J. Solid State Electrochem.* **10** 746–52

[12] Głab S, Hulanicki A, Edwall G and Ingman F 1989 Metal–metal oxide and metal oxide electrodes as pH sensors *Crit. Rev. Anal. Chem.* **21** 29–47

[13] Fog A and Buck R P 1984 Electronic semiconducting oxides as pH sensors *Sens. Actuators* **5** 137–46

[14] Khan M I, Mukherjee K, Shoukat R and Dong H 2017 A review on pH sensitive materials for sensors and detection methods *Microsyst. Technol.* **23** 4391–404

[15] Shahzad U, Saeed M, Marwani H M, Al-Humaidi J Y, Rehman S U, Althomali R H, Awual M R and Rahman M M 2025 Recent progress on potentiometric sensor applications based on nanoscale metal oxides: a comprehensive review *Crit. Rev. Anal. Chem.* **55** 1081–98

[16] Edwall G 1978 Improved antimony–antimony (III) oxide pH electrodes *Med. Biol. Eng. Comput.* **16** 661–9

[17] Horton B E, Schweitzer S, DeRouin A J and Ong K G 2010 A varactor-based, inductively coupled wireless pH sensor *IEEE Sens. J.* **11** 1061–6

[18] McMurray H N, Douglas P and Abbot D 1995 Novel thick-film pH sensors based on ruthenium dioxide–glass composites *Sens. Actuators B* **28** 9–15

[19] Trasatti S 1991 Physical electrochemistry of ceramic oxides *Electrochim. Acta.* **36** 225–41

[20] Mihell J and Atkinson J 1998 Planar thick-film pH electrodes based on ruthenium dioxide hydrate *Sens. Actuators B* **48** 505–11

[21] Liao Y-H and Chou J-C 2009 Preparation and characterization of the titanium dioxide thin films used for pH electrode and procaine drug sensor by sol–gel method *Mater. Chem. Phys.* **114** 542–8

[22] Al-Hilli S and Willander M 2009 The pH response and sensing mechanism of n-type ZnO/electrolyte interfaces *Sensors* **9** 7445–80

[23] Yates D E, Levine S and Healy T W 1974 Site-binding model of the electrical double layer at the oxide/water interface *J. Chem. Soc., Faraday Trans. 1 F* **70** 1807–18

[24] Kurzweil P 2009 Precious metal oxides for electrochemical energy converters: pseudocapacitance and pH dependence of redox processes *J. Power Sources* **190** 189–200

[25] Chen M, Jin Y, Qu X, Jin Q and Zhao J 2014 Electrochemical impedance spectroscopy study of Ta_2O_5 based EIOS pH sensors in acid environment *Sens. Actuators B* **192** 399–405

[26] Moss R E 2012 *Investigation of Porous Metal Oxide Coatings for a Novel Electrochemical Sensor for Orthophosphate* (University of Wisconsin-Madison)

[27] Lvovich V F 2012 *Impedance Spectroscopy: Applications to Electrochemical and Dielectric Phenomena* (New York: Wiley)

[28] Pejcic B and Marco R D 2006 Impedance spectroscopy: over 35 years of electrochemical sensor optimization *Electrochim. Acta.* **51** 6217–29

[29] Gill E, Arshak K, Arshak A and Korostynska O 2008 Mixed metal oxide films as pH sensing materials *Microsyst. Technol.* **14** 499–507

[30] Manjakkal L, Djurdjic E, Cvejin K, Kulawik J, Zaraska K and Szwagierczak D 2015 Electrochemical impedance spectroscopic analysis of RuO_2-based thick film pH sensors *Electrochim. Acta.* **168** 246–55

[31] Liao H-K, Chi L-L, Chou J-C, Chung W-Y, Sun T-P and Hsiung S-K 1999 Study on pH PZC and surface potential of tin oxide gate ISFET *Mater. Chem. Phys.* **59** 6–11

[32] Tombácz E 2009 pH-dependent surface charging of metal oxides *Period. Polytech. Chem. Eng.* **53** 77

[33] Al-Hilli S M, Willander M, Öst A and Strålfors P 2007 ZnO nanorods as an intracellular sensor for pH measurements *J. Appl. Phys.* **102** 084304

[34] Fulati A, Usman Ali S M, Riaz M, Amin G, Nur O and Willander M 2009 Miniaturized pH sensors based on zinc oxide nanotubes/nanorods *Sensors* **9** 8911–23

[35] Manjakkal L, Sakthivel B, Gopalakrishnan N and Dahiya R 2018 Printed flexible electrochemical pH sensors based on CuO nanorods *Sens. Actuators B* **263** 50–8

[36] Koncki R and Mascini M 1997 Screen-printed ruthenium dioxide electrodes for pH measurements *Anal. Chim. Acta.* **351** 143–9

[37] Chin Y-L, Chou J-C, Sun T-P, Liao H-K, Chung W-Y and Hsiung S-K 2001 A novel SnO_2/Al discrete gate ISFET pH sensor with CMOS standard process *Sens. Actuators B* **75** 36–42

[38] Kwon D-H, Cho B-W, Kim C-S and Sohn B-K 1996 Effects of heat treatment on Ta_2O_5 sensing membrane for low drift and high sensitivity pH-ISFET *Sens. Actuators* B **34** 441–5

[39] Huang W-D, Cao H, Deb S, Chiao M and Chiao J C 2011 A flexible pH sensor based on the iridium oxide sensing film *Sens. Actuators* A **169** 1–11

[40] Vijayakumar M, Pham Q N and Bohnke C 2005 Lithium lanthanum titanate ceramic as sensitive material for pH sensor: influence of synthesis methods and powder grains size *J. Eur. Ceram. Soc.* **25** 2973–6

[41] Martínez-Máñez R, Soto J, García-Breijo E, Gil L, Ibáñez J and Gadea E 2005 A multisensor in thick-film technology for water quality control *Sens. Actuators* A **120** 589–95

[42] Manjakkal L, Zaraska K, Cvejin K, Kulawik J and Szwagierczak D 2016 Potentiometric RuO_2–Ta_2O_5 pH sensors fabricated using thick film and LTCC technologies *Talanta* **147** 233–40

[43] Zhuiykov S 2009 Morphology of Pt-doped nanofabricated RuO_2 sensing electrodes and their properties in water quality monitoring sensors *Sens. Actuators* B **136** 248–56

[44] Labrador R H, Soto J, Martínez-Máñez R, Coll C, Benito A, Ibáñez J, García-Breijo E and Gil L 2007 An electrochemical characterization of thick-film electrodes based on RuO_2-containing resistive pastes *J. Electroanal. Chem.* **611** 175–80

[45] Manjakkal L, Cvejin K, Kulawik J, Zaraska K, Szwagierczak D and Socha R P 2014 Fabrication of thick film sensitive RuO_2–TiO_2 and Ag/AgCl/KCl reference electrodes and their application for pH measurements *Sens. Actuators* B **204** 57–67

[46] Liao Y-H and Chou J-C 2008 Preparation and characteristics of ruthenium dioxide for pH array sensors with real-time measurement system *Sens. Actuators* B **128** 603–12

[47] Tsai C N, Chou J C, Sun T P and Hsiung S K 2006 Study on the time-dependent slow response of the tin oxide pH electrode *IEEE Sens. J.* **6** 1243–9

[48] Chen P-Y, Yin L-T, Shi M-D and Lee Y-C 2013 Drift and light characteristics of EGFET based on SnO_2/ITO sensing gate *Life Sci. J.* **10** 3132–6

[49] Chou J C and Liao L P 2004 Study of TiO_2 thin films for ion sensitive field effect transistor application with RF sputtering deposition *Jpn. J. Appl. Phys.* **43** 61

[50] Pocrifka L, Goncalves C, Grossi P, Colpa P and Pereira E 2006 Development of RuO_2–TiO_2 (70–30) mol% for pH measurements *Sens. Actuators* B **113** 1012–6

[51] Da Silva G, Lemos S, Pocrifka L, Marreto P, Rosario A and Pereira E 2008 Development of low-cost metal oxide pH electrodes based on the polymeric precursor method *Anal. Chim. Acta.* **616** 36–41

[52] Chou J C and Wang Y F 2002 Preparation and study on the drift and hysteresis properties of the tin oxide gate ISFET by the sol–gel method *Sens. Actuators* B **86** 58–62

[53] Marzouk S A, Ufer S, Buck R P, Johnson T A, Dunlap L A and Cascio W E 1998 Electrodeposited iridium oxide pH electrode for measurement of extracellular myocardial acidosis during acute ischemia *Anal. Chem.* **70** 5054–61

[54] Prats-Alfonso E, Abad L, Casañ-Pastor N, Gonzalo-Ruiz J and Baldrich E 2013 Iridium oxide pH sensor for biomedical applications. Case urea–urease in real urine samples *Biosens. Bioelectron.* **39** 163–9

[55] Santos L, Neto J P, Crespo A, Nunes D, Costa N, Fonseca I M, Barquinha P, Pereira L, Silva J and Martins R 2014 WO_3 nanoparticle-based conformable pH sensor *ACS Appl. Mater. Interfaces* **6** 12226–34

Chapter 5

Classification of metal oxides for electrochemical pH sensors fabrication and their performance

5.1 Introduction

As discussed in the previous chapter, metal oxides (MOx) have attracted significant interest for developing electrochemical pH sensors. The MOx were used in all types of electrochemical pH sensors, and in potentiometric sensors received greater attention due to their performance. It was found that the sensitivity of MOx-based pH sensors depends on both the type of materials used for their fabrication and the fabrication method. The previous review papers compared the sensing performances of various materials and methods of fabrication to evaluate the best sensitivity of the materials [1–3]. The ionic and electronic conductivity of the materials is an important factor in enhancing the performance of MOx-based pH sensors. It was noticed that MOx have mixed ionic and electronic conductivity. To improve the Nernstian response of the sensors, researchers have developed single MOx- and mixed oxides-based pH sensors [4–6]. The electrochemical performances reveal that the sensor exhibits sub-Nernstian, Nernstian and super-Nernstian responses depending on the type of materials and method of fabrication [2–4]. The Nernstian response of a few MOx were considered, such as IrO_2 [7–11], RuO_2 [7, 12–15], TiO_2 [7, 16–19], SnO_2 [7, 20–23], Ta_2O_5 [7, 24–27], WO_3 [28–32] and ZnO [33–37]. The observed sensitivity compared with the theoretical sensitivity at 25 °C derived from the Nernst equation (59.14 mV/pH). The observed sensitivity variation is shown in figure 5.1, and it is found that the RuO_2-based pH sensors exhibit a stable pH sensing performance [2]. This chapter discusses the sensing mechanism of various classes of metal oxides and compares their performances.

doi:10.1088/978-0-7503-6079-1ch5

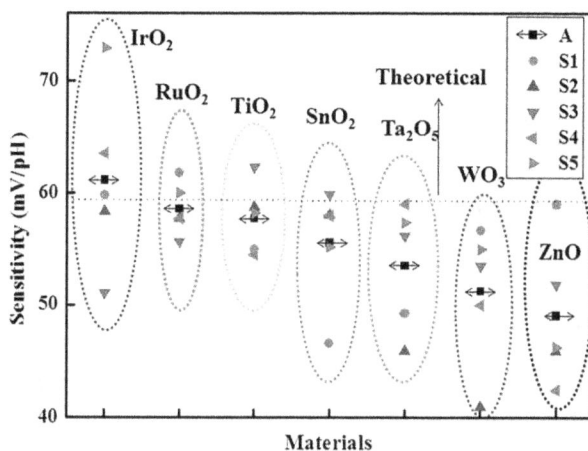

Figure 5.1. Comparison of pH sensitivity of various metal oxides with theoretical value. Reproduced from [2] CC BY 4.0.

5.2 Different metal oxides for pH sensing

5.2.1 Ruthenium (IV) oxide (RuO₂) based pH sensor

RuO_2 belongs to electrically conducting transition metal oxide compounds with a rutile structure and a space group D_{4h}^{14}. According to the factor group analysis, RuO_2 consists of fifteen optical phonon modes of irreducible representation. The major Raman active modes for RuO_2 are A_{1g}, B_{1g}, B_{2g} and E_g. The intensity of the soft phonon mode B_{1g} is very weak, and it is normally not observed for the film [38, 39]. Figure 5.2(a) shows the spatial arrangement of Ru and O atoms in the rutile structure and the corresponding three principle Raman active modes E_g, A_{1g} and B_{2g} [38, 39]. For the investigated RuO_2 thick films, three Raman active modes were found with positions: $E_g = 526$ cm^{-1}, $A_{1g} = 644$ cm^{-1} and $B_{2g} = 714$ cm^{-1}, shown in figure 5.2(b). The doubly degenerated mode E_g has higher intensity as compared with other singlet modes (A_{1g} and B_{2g}) and the position of the peak corresponds to the single crystal of RuO_2 in the (101) orientation [38, 39]. RuO_2 is a promising material used for making pH sensors both potentiometric and conductimetric (as shown in figures 5.2(c) and (d)) due to its chemical stability and high conductivity, which inhibits the space charge accumulation. For pH sensor fabrication, either pristine RuO_2 or combined with other oxides are used [5, 40, 41]. Several methods have been employed for RuO_2-based pH sensor fabrication, including thin- and thick-film technologies.

According to Fog and Buck [7], there are two main reasons for pH response in RuO_2 metal oxide: (i) 'oxygen intercalation', and (ii) ion exchange in a surface layer containing –OH groups. The general mechanism was explained as:

(i) For oxygen intercalation

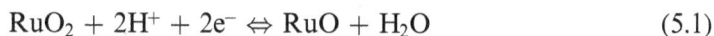

$$RuO_2 + 2H^+ + 2e^- \Leftrightarrow RuO + H_2O \qquad (5.1)$$

where RuO_2 and RuO are higher and lower metal oxide.

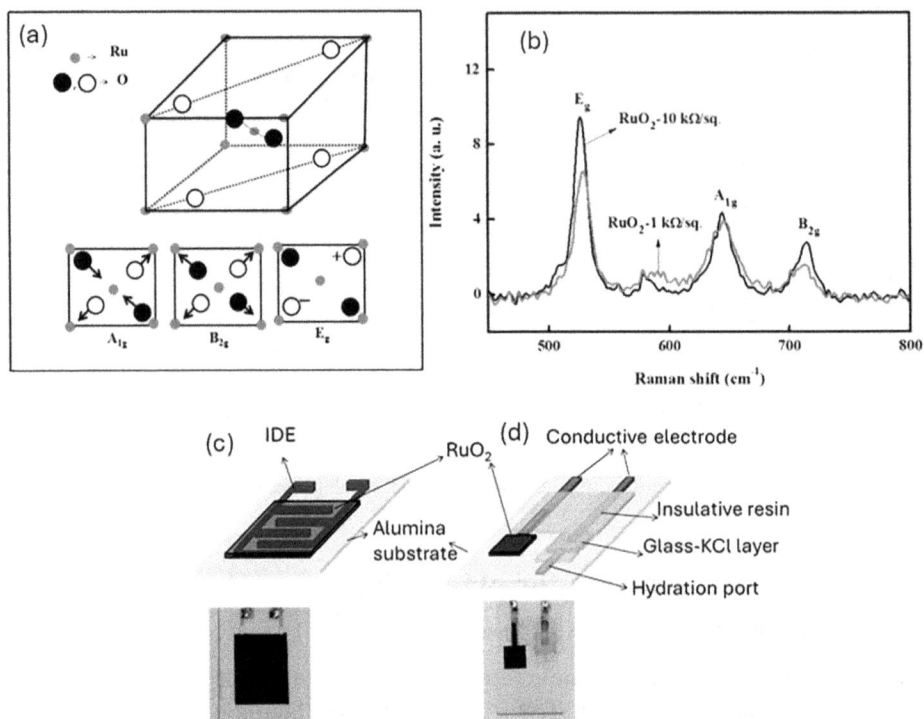

Figure 5.2. (a) Spatial arrangements of atoms in the RuO_2 rutile structure and the corresponding Raman active modes. Reprinted from [6], copyright (2015), with permission from Elsevier. (b) Raman spectra of two RuO_2-based thick films printed on alumina substrate. Schematic view and microscopic images of the RuO_2-based pH sensors, reprinted from [94], copyright (2015), with permission from Elsevier, (c) conductimetric, (d) potentiometric, reproduced from [2] CC BY 4.0.

(ii) For proton exchange

$$RuO_x(OH)_y + \delta H^+ + \delta e^- \Leftrightarrow RuO_{x-\delta}(OH)_{y+\delta} \tag{5.2}$$

McMurray *et al* [42] and Trasatti [43] also confirmed this. As per previous reports [3, 44] for a potentiometric sensor, in acidic solution, the proton H^+ is released and the potential increases with rising pH towards acidic [3, 44].

$$2[Ru^{III}]OH \Leftrightarrow 2[Ru^{IV}]O^{II} + 2H^+ + 2e^- \tag{5.3}$$

In a basic solution, the proton binds to $[Ru^{III}]$ clusters and the electrode potential decreases with rising pH towards a basic solution [3, 44].

$$2[Ru^{IV}]O^{II} + H_2O + 2e^- \Leftrightarrow 2[Ru^{III}]O + 2OH^- \tag{5.4}$$

The Nernstian response for Ru^{IV}/Ru^{III} reduction potential can be expressed as [3, 44]

$$E = E^0 - \ln \frac{RT}{F}\left[\text{pH} + \log \frac{a[\text{Ru}^{III}]}{a[\text{Ru}^{IV}]}\right] \qquad (5.5)$$

where: E_0—standard potential, R—gas constant, T—absolute temperature, F—Faraday constant, $a[\text{Ru}^{III}]$, $a[\text{Ru}^{IV}]$—activity of Ru ions at 3+ and 4+ oxidation state, respectively.

Hence, for the fabricated thick-film pH sensor, the potential difference generated between the electrodes depends on the concentration of H^+ in the solution. In 1995, McMurray et al studied $RuO_2 \cdot xH_2O$: lead borosilicate glass composite as a pH sensing electrode [42]. They noted that higher glass content in the composite causes a less open pore structure, which leads to a decrease in the diffusion rate of the proton to RuO_2 surface, resulting in a decrease in the response time of the MOx electrodes. They also found that when the sensing electrode is incorporated with a platinum wire, the response slope is higher than the theoretical value, which may be due to the electrical characteristics of the platinum RuO_2-glass composite contact. The authors explained that the response to pH changes can be affected by the storage environment of the sensitive electrode [42]. The storage medium of the developed electrodes is also strongly influencing the sensing performance [45]. For example, before electrochemical measurement, if the electrode is dipped in deionised water, the stability and sensing performance of the RuO_2-based pH sensor will be enhanced [46]. The previous study of Manjakkal et al found observed that the effect of sheet resistivity and storing conditions also influenced the determination of pH in a RuO_2-based pH sensor [47–49]. Changing pH response while storing is possibly due to the interaction between protons and hydroxide groups on the oxide surface.

To reduce the cost of the RuO_2 sensitive electrode and enhance the performances of the sensor, the doping or mixing of other oxides into RuO_2 were implemented. A nanostructured Pt-doped RuO_2-based pH sensor shows an excellent Nernstian response with a very fast response time [46]. A comparison of the performance of sensitivity is shown in figure 5.3(a) [46]. In addition, it was found that the mixed

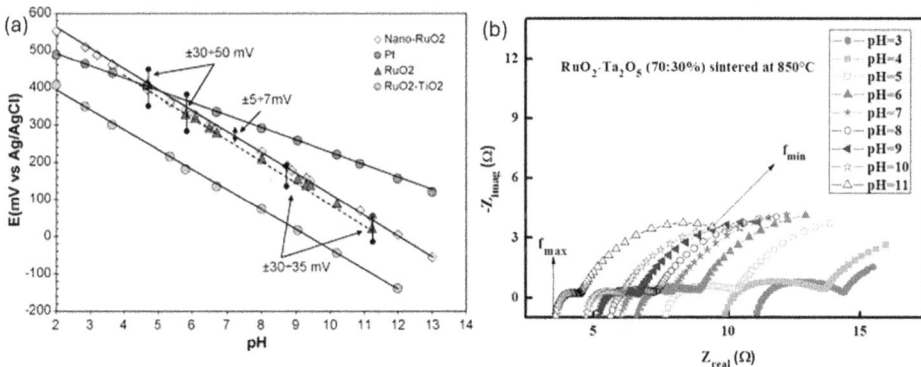

Figure 5.3. (a) Comparison of potentiometric performances of various electrodes. Reprinted from [46], copyright (2016), with permission from Elsevier. (b) EIS analysis of RuO_2–Ta_2O_5-based pH sensing electrode. Reprinted from [52], copyright (2011), with permission from Elsevier.

oxide composition helps reduce the cost of the sensing electrode, and the sensor shows good stability and activity [5, 50, 51]. Pocrifka *et al* [51] developed a RuO_2–TiO_2 binary metal oxide-based pH-sensitive electrode by the Pechini method. The fabricated sensor shows a sensitivity of 56.03 mV/pH [51]. As a case study, Manjakkal previously worked with many binary oxides of RuO_2 with SnO_2, TiO_2 and Ta_2O_5 for the design of potentiometric and conductimetric pH sensor fabrication. The performances of these sensors are given in table 5.1.

The sensor shows a sensitivity close to Nernstian response with a very fast response time. It was observed that the combined ionic and electronic properties involved in the pH sensing mechanism are observed through the electrochemical

Table 5.1. Potentiometric performance of thick-film pH sensors with RuO_2-based binary oxides and single RuO_2-based electrodes [5, 6, 47–50].

Properties	RuO_2–Ta_2O_5 (70:30)wt.%	RuO_2–SnO_2 (70:30)wt.%	RuO_2–TiO_2 (70:30)wt.%	RuO_2–Ta_2O_5–glass	RuO₂ ESL pastes 1 kΩ sq⁻¹	10 kΩ sq⁻¹
Sensitivity (mV/pH)	56.4	56.5	56.1	58.0	60.0	60.1
Correlation coefficient	0.999	0.998	0.998	0.999	0.978	0.994
Response time	<8 s in acidic solutions <15 s in basic solutions	<15 s	<15 s	<15 s	<60 s	<60 s
Long-term stability (stored in atmospheric conditions)	Very good (after 1 year, sensitivity 55 mV/pH)	Very good (after 6 months sensitivity 56 mV/pH)	Very good (after 5 months, sensitivity change of ±0.5%)	Very good (small sensitivity changes after 2 months)	Very good	Very good
Hysteresis effect	±3 mV in acidic, ±8 mV in basic solutions	±2 mV in acidic, ±7 mV in basic solutions	±3 mV in acidic, ±5 mV in basic solutions	±3 mV in acidic, ±10 mV in basic solutions	—	—
Drift effect	Very low	Very low	Very low	Very low	Low	Low
Influence of solution conductivity (>2500 µS)	8 mV potential change	2 mV potential change	5 mV potential change	—	—	—
Interference to other ions	Very low	Very low	Very low	Very low	Low	Low

impedance spectroscopy (EIS) analysis. As an example, for RuO_2–Ta_2O_5 composites figure 5.3(b) displays the Nyquist plot in the frequency range 10 Hz–2 MHz at various pH values of the solution [52]. Three regions in increasing frequency order can be distinguished in this plot, corresponding to the dominant role of ion diffusion, charge transfer resistance R_{ct} and solution resistance R_s [52]. A smaller arc in the higher frequency range is attributed to charge transfer phenomena at the electrode–solution interface, while an incomplete, larger arc at low frequency is due to ion diffusion processes in the interface region. It was stated that the frequency corresponding to the maximum of low frequency arc increases with rising pH, thus reflecting faster relaxation of diffusion processes in alkaline than in acidic solutions. The charge transfer resistance R_{ct} is reflected by the diameter of the smaller semicircle. The frequency corresponding to this arc is almost independent of pH. The nonzero intersection of this arc with the real axis at the high frequency side is assigned to the solution resistance R_s and other minor ohmic resistances, like oxide film resistance [52]. Due to the prevailing content of highly conductive RuO_2 phase, the contribution of the resistance of the sensing oxide layer is predicted to be very small for the RuO_2–Ta_2O_5 (70:30%) sensor. It was observed that in the high frequency range, the solution resistance has a strong influence on the real part of the sensor impedance and decreases with rising pH [52]. Table 5.2 represents the methods of fabrication and properties of a few reported RuO_2-based pH sensors.

5.2.2 Iridium oxide (IrO_2) based pH sensor

Iridium oxide (IrO_x)-based pH sensors have received significant attention in recent years due to their excellent electrochemical sensing performance [53]. In the class of MOx-based pH sensors, IrO_x has shown several advantages, including good stability over a wide pH range, even at high temperature up to 250 °C, at high pressure, and in an aggressive environment, with fast response time even in non-aqueous solutions [54, 55]. Like RuO_2, IrO_2 also crystallises in the rutile structure [56]. Compared with RuO_2, even though the Ir salts are more expensive than Ru salts, the studies reveal that the IrO_2 has less catalytic activity than the RuO_2. Moreover, its application as a pH sensor, IrOx is also well known, as an anode material for oxygen and chlorine evolution and as an electrochromic material [57, 58]. Three possible redox reactions were involved for pH sensing of the iridium oxides as [9]

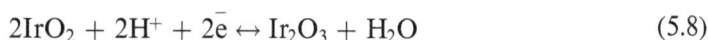

$$Ir_2O_3 + 6H^+ + 6\bar{e} \leftrightarrow 2Ir + 3H_2O \tag{5.6}$$

$$IrO_2 + 4H^+ + 4\bar{e} \leftrightarrow Ir + 2H_2O \tag{5.7}$$

$$2IrO_2 + 2H^+ + 2\bar{e} \leftrightarrow Ir_2O_3 + H_2O \tag{5.8}$$

The redox potential is determined by using Nernstian equation. For example, a flexible IrO_2-based pH sensor fabricated using sol–gel process, as shown in figures 5.4 (a) and (b) [9]. The sol–gel method of fabrication of IrO_2 film could provide a simpler and economically feasible approach. This process is also helpful for the low-temperature heat treatment of the film. By using the fabricated sensor, three tests

Table 5.2. Comparison of performances of various MOx-based pH sensors.

Material	Fabrication methods	Properties	References
RuO_2 nanoparticles-modified vertically aligned carbon nanotubes (RuO_2/MWCNTs)	RuO_2 was deposited by magnetron sputtering on the vertically aligned MWCNTs which were grown on Ta substrates	• Sensitivity −55 mV/pH from pH 2 to 12. • Hysteric width for loop cycle pH 7–4–7–10–7 is 6.4 mV, pH 7–10–7–4–7 is 5.1 mV and for pH 2–8–12–8–2 is 10.2 mV. • Response time less than 40 s, good reproducibility, long-term storage stability (over 1 month), negligible interference of ions in the pH measurement. • Charge transfer resistance R_{ct} gradually increase from 1.80 to 10.5 kΩ from pH 2 to 12.	[76]
Pt-doped RuO_2	Screen printing	• Nernstian slope was 58 mV/pH in the range of pH 2–13. • Excellent response time 1–2 s, reproducibly and long-term potential stability (8 months). • The sensitive electrode was also capable for measuring dissolved oxygen (DO) in the range 0.6–8.00 ppm.	[46]
Graphite based ink with 9.8% RuO_2	Screen printing	• Disposable pH sensor. • Sensitivity 51 mV/pH. • Linear response up to pH 8. • Potential application in some natural drinks.	[77]
RuO_2	Radio frequency sputtering	• Sensitivity 55.64 mV/pH from pH 1 to 13. • Drift rate for pH 4 is 0.13 mV/pH, pH 7 is 0.38 mV/pH and for pH 10 is 7.31 mV/pH.	[13]

(Continued)

Table 5.2. (*Continued*)

Material	Fabrication methods	Properties	References
		• Hysteresis width for loop pH 7-4-7-10-7 is 4.36 mV and for loop pH 7-10-7-4-7 is 2.2 mV. • Average selectivity coefficient of the sensor for ion K^+ is −5.38 and for Na^+ is −5.43. • Used for real-time pH measurement system.	
$RuO_2 \cdot xH_2O$ mixed with polymer paste in the ratio of is 1:2 by weight	Screen printing	• Sensitivity 52.1 mV/pH in the range of pH 2–10. • Sensitivity is dependent on storage conditions.	[45]
$RuO_2 \cdot xH_2O$: glass mixture	Hand painting	• Observed sensitivity is 59 mV/pH in the range 2–12, and it depends on the oxide: glass composition and fabrication steps.	[42]
RuO_2–TiO_2	Pechini method	• Sensitivity 56.03 mV/pH in the range of pH 2–12. • Insensible the presence of Na^+, K^+ and Li^+ ions in the solution.	[78]
Ni-RuO_2	Thermal decomposition	• Sensitivity 52 ± 2 mV/pH in the range of pH 1.5–12.5. • Excellent reproducibility and stability. • Only small interference for some cations and anions.	[79]
Cu_2O doped RuO_2	Screen printing	• Response time of planar pH sensor improved from 80–120 to 25 s with increasing thickness from 2 to 5 μm.	[80]
RuO_2–TiO_2	Screen printing	• Sensitivity 56.11 mV/pH in the range of pH 2–12. • Response time less than 15 s.	[5]

Material	Fabrication	Remarks	Ref.
		• Excellent stability and long lifetime. • EIS analysis of the sensitive electrode was monitored. • XPS analysis explains the formation of surface hydroxyl groups and the change of binding energy for a selected pH solution.	
$Bi_2Ru_2O_{7+x}$–RuO_2	Screen printing	• Sensitivity of 58 mV/pH in the pH range 2–13 • Sensor shows a long lifetime with 18 months are testing. • The selectivity coefficient of interference of other cations and anions is lower. • Standard deviation in the output electromotive force was ±7 mV. • Sensor can detect dissolved oxygen concentration in the range of 0.5–8.00 ppm.	[81]
RuO_2-comercial paste	Screen printing	• Sensitivity 57 mV/pH in the range of pH unit 4–11. • The theoretical model was explained for determining the interference of anions.	[82]
RuO_2-comercial paste	Screen printing	• Sensitivity 57 ± 3 mV/pH up to 12 units of pH. • Response time is lower than 5 s, and lifetime is greater than 6 months.	[12]
RuO_2	Screen printing	• Sensitivity of 53 mV/pH (±%) was obtained for the sensitive electrode with a thick-film polymer-based reference electrode in the pH range 4 to 10.	[83]

(Continued)

Table 5.2. (*Continued*)

Material	Fabrication methods	Properties	References
RuO$_2$-Commercial pastes (RuO$_2$–1 kΩ sq^{-1} and RuO$_2$–10 kΩ sq^{-1})	Screen printing	• Sensitivity of 50 mV/pH ($\pm$2.5%) was obtained for the sensitive electrode with an external commercial saturated calomel reference electrode. • Polymer-based reference electrode sensors show a small drift in potential during 6 months of testing. • Sensitivity of RuO$_2$–1 kΩ sq^{-1} is 60.01 mV/pH and for RuO$_2$–10 kΩ sq^{-1} is 60.78 mV/pH in the range of pH 2–10. • RuO$_2$–1 kΩ sq^{-1}-based pH sensor shows faster response than RuO$_2$–10 kΩ sq^{-1}-based pH sensor. • Suitable for conductimetric pH sensor fabrication.	[47–49]
IrO$_2$	Sol–gel-dip coating	• A sensitivity repeatedly and reversibly between 51.1 and 51.7 mV/pH in the pH range between 1.5 °C and 12 at 25 °C was observed.	[9]
IrO$_2$	Carbonate melt oxidation method	• Sensitivity 58.4 $\pm$ 0.2 mV/pH in the pH range 1–13. • Excellent stability over a long period of 2.5 years.	[8]
IrO$_2$–TiO$_2$	Polymeric precursor method	• Sensitivity of IrO$_2$–TiO$_2$ (100–0) mol% is 69.9 $\pm$ 3.66 mV/pH, IrO$_2$–TiO$_2$ (70–30) mol% is 58.7 $\pm$ 2.34 mV/pH, IrO$_2$–TiO$_2$ (30–70) mol% is 59.1 $\pm$ 1.47 mV/pH and IrO$_2$–TiO$_2$ (20–80) mol% is 61.2 $\pm$ 1.99 mV/pH.	[84]

		• IrO_2–TiO_2 (30–70) mol% in the range of pH 1–13 shows excellent sensitivity, fast response, good reproducibility. • Interfering ions susch susch as K^+, Na^+ and Li^+ on the sensitive electrode is not significant.	
IrO_2	Electrodeposition method	• Sensor shows a super-Nernstian response with a sensitivity 72.9 ± 0.9 mV/pH in the range of pH 3–11. • Sensor provides a detection of 0.1–20 mM of urea in just 50 μl.	[11]
IrO_2	Anodic electrodeposition method	• A super-Nernstian response with a sensitivity of 63.5 ± 2.2 mV/pH in the pH range 2–10. • Lifetime of 1 month with an accuracy of about 0.02 pH. • The hysteresis of the potential was very small, nearly 2.5 ± 0.6 mV and the interference of ions was very less.	[10]
SnO_2	Sputtering	• Sensitivity 58 mV/pH in the pH range 2–10.	[22]
SnO_2	Sputtering	• Sensitivity 59.17 mV/pH in the pH range 2–12.	[85]
Ta_2O_5	Sputtering	• Sensitivity (58–59) mV/pH in the pH range 2–12. • Long-term drift (0.03–0.05 pH/day and fast response less than 0.3 s.	[26]

(Continued)

Table 5.2. (*Continued*)

Material	Fabrication methods	Properties	References
Ta_2O_5—EIS (Electrolyte-Insulator-Semiconductor) Sensor	Electron beam evaporation	• Sensitivity 57 ± 1.5 mV/pH in the pH range 2–12. • The drift of the output signal after 3 and 10 h for pH = 7 is less than 0.01 pH/h and 0.004 pH/h, respectively and lower hysteresis less than 1 mV for pH 0.02.	[86]
Ta_2O_5	Screen printing	• Sub-Nernstian (45 mV/pH) response was found in the pH range 2–10. • The conductance, capacitance and impedance of conductimetric sensor are strongly dependent on pH in the low frequency range.	[24]
Ta_2O_5—EIOS pH sensor	Sputtering and thermal oxidation	• Nernstian pH sensitivity 56.19 mV/pH was observed in the pH range 1–10. • Impedance parameters of the sensor do not change much with changing pH of the solution.	[87]
Al_2O_3–Ta_2O_5 and Al_2O_3–ZrO_2	Metal oxide chemical vapour deposition (MOCVD)	• The sensitivity for Al_2O_3 is 51.5 mV/pH in the range of pH 0.5–11, for Ta_2O_5 is 56.2 mV/pH in the range of pH 2–11.5, and for ZrO_2 is 51.0 mV/pH in the range of pH 0–12. • The sensitivity of Al_2O_3–Ta_2O_5 with mmole fraction of Ta $(X_{Ta}) = 0.63$ is 56.7 mV/pH and that for Al_2O_3–ZrO_2 with a mole fraction of Zr $(X_{Zr}) = 0.63$ is 52.3 mV/pH.	[88]
CeO_2 $Ce_{0.8}Sm_{0.2}O_2$	Screen printing	• The average sensitivity of CeO_2 is 38 ± 4 mV/pH, $Ce_{0.8}Sm_{0.2}O_2$ is 40 ± 4 mV/pH and for	[89]

Material	Preparation method	Description	Ref.
$Ce_{0.8}Zr_{0.2}O_2$		$Ce_{0.8}Zr_{0.2}O_2$ 51 ± 2 mV/pH in the pH range 7.2–10.8.	[90]
Cobalt oxide	Screen printing	• The pH sensitivity depends on the preparation method of cobalt oxide. • The Co content in the paste (5%–10%) shows excellent sensitivity in the range 54.9 to 60.3 mV/pH in the pH range 1–12. • Response time is less than 1 min, and there is no hysteresis effect.	[90]
WO$_3$/MWNT	Magnetron sputtering	• Sensitivity 41 mV/pH was observed in the pH range 2–12. • Response time less than 90 s, stability over one month and good reproducibility.	[29]
W/WO$_3$	Electrooxidation	• Sensitivity of the sensor is 53.5 ± 0.5 mV/pH in the pH range 2–12. • For the pH 6–7 it shows a response time of 3 s, but for high pH the response time is 15 s.	[30]
WO$_3$	RF sputtering	• Sensitivity of the sensor under different temperatures, 25 °C, 35 °C, 45 °C, 55 °C, and 65 °C, is 44.85, 47.90, 50.10, 52.85 and 55.80 mV/pH, respectively. • The hysteresis width for the loop pH = 3–1–3–5–3 is 7.2 mV, pH = 3–5–3–1–3 is 12.5 mV • pH = 4–1–4–7–4 is 12 mV, pH = 4–7–4–1–4 is 26 mV, and the hysteresis width for the loop pH 3–1–	[65]

(Continued)

Table 5.2. (*Continued*)

Material	Fabrication methods	Properties	References
		3–5–3 at various loop times 600, 1200 and 2400 s are 7.2, 9.7 and 15.4 mV, respectively. Also, for the loop's pH = 3–1–3–5–3 and pH = 4–1–4–7–4 at 600 s is 7.2 and 12 mV, respectively. • The drift rate for pH = 1, 3, 5, 7 are 1.5, 3.6, 6.6, 15.7 mV h^{-1}, respectively.	
PbO$_2$	Al electrode dipping into PbO$_2$ solution and transformed into α-PbO$_2$ and β-PbO$_2$ by its oxidation at alkaline and acidic solutions	• Sensitivity of the sensor based on α-PbO$_2$ and β-PbO$_2$-modified electrodes is 57.96 and 57.80 mV/pH, respectively in the pH range 1–12.	[69]
PbO$_2$	Electrodeposition	• Sensitivity of the sensor based on α-PbO$_2$ and β-PbO$_2$ is 64.82 mV/pH and 57.85 mV/pH, respectively, in the range of pH 1.5–12.5. • The response time of the sensor in acidic solution is 1 s and in basic solution is 30 s.	[70]
LLTO	Sol–gel and solid-state methods	• In the solid-state reaction method of preparation, the sensor shows a sensitivity of (40–50) mV/pH. • In the sol–gel method of preparation of the electrode, the sensor gives a sensitivity of 42 mV/pH.	[91]
TiO$_2$	Sol–gel method was spin-coated on the ITO substrate	• Sensitivity of the pH electrode was 58.75 mV/pH in the range of pH between 1 and 11. • The drift rate of the sensor was 1.97 mV h^{-1} in pH buffer solution.	[16]

TiO$_2$ nanotube array modified Ti electrode	Anodization of Ti substrate electrode	• The sensitivity of the electrode is 54.6 mV/pH from pH 2 to 12 and 57.1 mV/pH for 12–2, with a response time of less than 30 s. • The sensitivity was enhanced 59.3 mV/pH from 2 to 12 and 59.1 mV/pH from 12 to 2, respectively, after being irradiated with UV rays. • The interference from the most common ions was negligible.	[92]
TiO$_2$/MWCNT/cellulose hybrid Nanocomposite	TiO$_2$/MWCNT blended cellulose solution coated on a silicon wafer by a spin coater	• The conductimetric pH sensor exhibits two linear conductance in pH 1–12. • Highly sensitive in the region of pH 1–6 and less in the region 7–12.	[93]

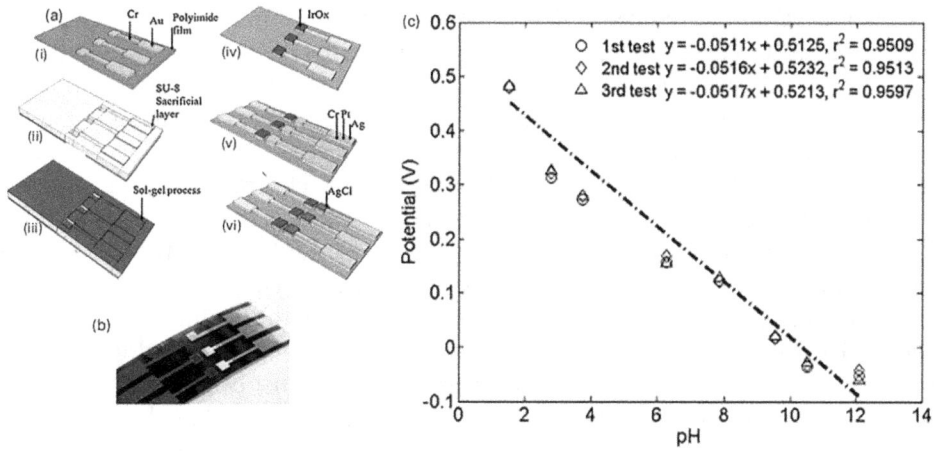

Figure 5.4. (a) Fabrication steps of a flexible IrO$_x$-based pH sensor. (b) Image of the fabricated flexible pH sensor array. (c) Open circuit potential for various tests of the developed flexible sensor. Reprinted from [9], copyright (2011), with permission from Elsevier.

were carried out, and it was found that the sensitivity is almost similar and is 51 mV/pH, as shown in figure 5.4(c) [9]. The results reveal that there is a variation in the standard potential (E_0) of the sensor. The variation could be due to the difference in oxidation state and also the ion exchange of the surface of the IrO$_x$ film containing OH-groups. The general equilibrium of the electrodes is described as [9]

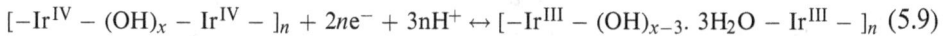

$$[-Ir^{IV} - (OH)_x - Ir^{IV} -]_n + 2ne^- + 3nH^+ \leftrightarrow [-Ir^{III} - (OH)_{x-3}. 3H_2O - Ir^{III} -]_n \quad (5.9)$$

$$Ir_2O(OH)_3O_3^{3-} + 3H^+ + 2e^- \leftrightarrow 2IR(OH)_2O^- + H_2O \quad (5.10)$$

As per this equilibrium, due to more electrodes produced, the sensitivity may be different from the theoretical values [9]. In addition to this, the characteristics of IrO$_x$ sensors are generally very sensitive to their structure and composition, which depend on the fabrication methods and conditions [9]. For a long-term stable IrO$_2$ pH-sensitive electrode development, the chemical oxidation method called 'carbonate melt oxidation' shows a potential approach [8]. As proof, a uniform IrO$_2$ film is coated on the surface of an iridium metal wire through oxidation of the wire in a carbonate melt. This pH sensor responds over a wide pH range with a slope of 58.4 ± 0.2 mV/pH over a long period of 2.5 years. Moreover, it is suited for continuous pH measurement without the need for frequent calibration [8].In one of the recent studies, through the micro-electro-mechanical system (MEMS), a batch preparation of pH sensor is reported in which the IrO$_x$ electrode is prepared by electrodeposition and a solid Ag/AgCl with liquid KCl storage-based reference electrode developed [59]. The fabrication process of the sensor describes procedures for the development of both sensitive and reference electrodes. The response of the reference electrode exhibits a Nernstian behaviour with a sensitivity of the Ag/AgCl for Cl—sensing is around 57 mV/10 M and is close to the theoretical value. The integrated reference and sensing electrode exhibit a super-Nernstian response (82.88 mV/pH) [59].

Kinlen *et al* [60] developed a new approach for the fabrication of a pH sensor. They designed a pH sensor based on a Nafion-coated IrO_2 electrode and a polymer-based AgCl reference electrode. When this Nafion is coated on the IrO_2 surface, it becomes permselective, passing only cations and not anions. By maintaining the electrode sensitivity to pH, it eliminates or attenuates many of these interferences. The authors reported that the polymer-based reference electrode helped to trap the Cl^- ion in the polymer layer by encapsulating it with a Nafion outer layer. After annealing, the Nafion membrane effectively blocks Cl^- diffusion to the test solution and maintains a constant Cl^- activity on the AgCl surface, and thus a constant electrode potential. Hence, by applying this method, one of the drawbacks of metal oxide pH sensors is due to its electronic conduction it responds to redox species in solution, which leads to errors in pH measurement and must be avoided [60]. Based on different fabrication approaches, there are various IrO_2-based pH-sensitive electrodes developed and implemented for various biomedical and pollution-monitoring applications. The performance of a few sensors is mentioned in table 5.2.

5.2.3 WO_3 based pH sensor

WO_3 is considered an excellent 2D semiconductor which shows applications in sensing, catalysts and chromogenic materials [61–63]. The promising properties of this material, including good morphological and structural control, high sensitivity and selectivity, biocompatibility and structural control for nanomaterial formation, lead to them being a promising material for electrochemical sensors [61]. Furthermore, these materials are widely available and low-cost for large-scale production. WO_3-based sensitive materials are mainly used for microelectrochemical pH sensor development and have found applications in cell determination. For example, a W/WO_3-based ultra-micro pH sensor was developed for the monitoring of pH value from cultured endothelial cells and observing the pH variation generated from normal, damaged and recovery endothelial cells [30]. As a pH sensing application, this material is initially considered as an active electrode in a microelectrochemical transistor [64], and the configuration is shown in figure 5.5(a) [61]. In this microelectrochemical transistor, the sensor was operating either electrically or chemically by changing the gate voltage or the pH value of the solution [2]. The general mechanism for WO_3 as a proton-dependent redox reaction is shown below [64].

$$nM^+ + ne^- + WO_3 \Leftrightarrow M_nWO_3 \tag{5.11}$$

where M is usually a hydrogen ion but may also be a metal ion (Na^+, Li^+). According to Fog and Buck [7], the single phase 'oxygen intercalation', by omitting hydration, the surface mechanism can be expressed as

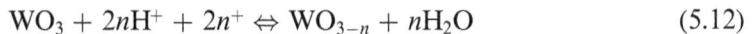

$$WO_3 + 2nH^+ + 2n^+ \Leftrightarrow WO_{3-n} + nH_2O \tag{5.12}$$

Then the electrode potential by the Nernst equation states that

$$E = E_0 - \frac{RT}{zF} \ln \frac{a_{(WO_{3-n})}}{a_{(WO_3)}} - \frac{RT}{F} \ln \frac{1}{a_{(H^+)}} = E_0 - \frac{2.303RT}{F}pH \tag{5.13}$$

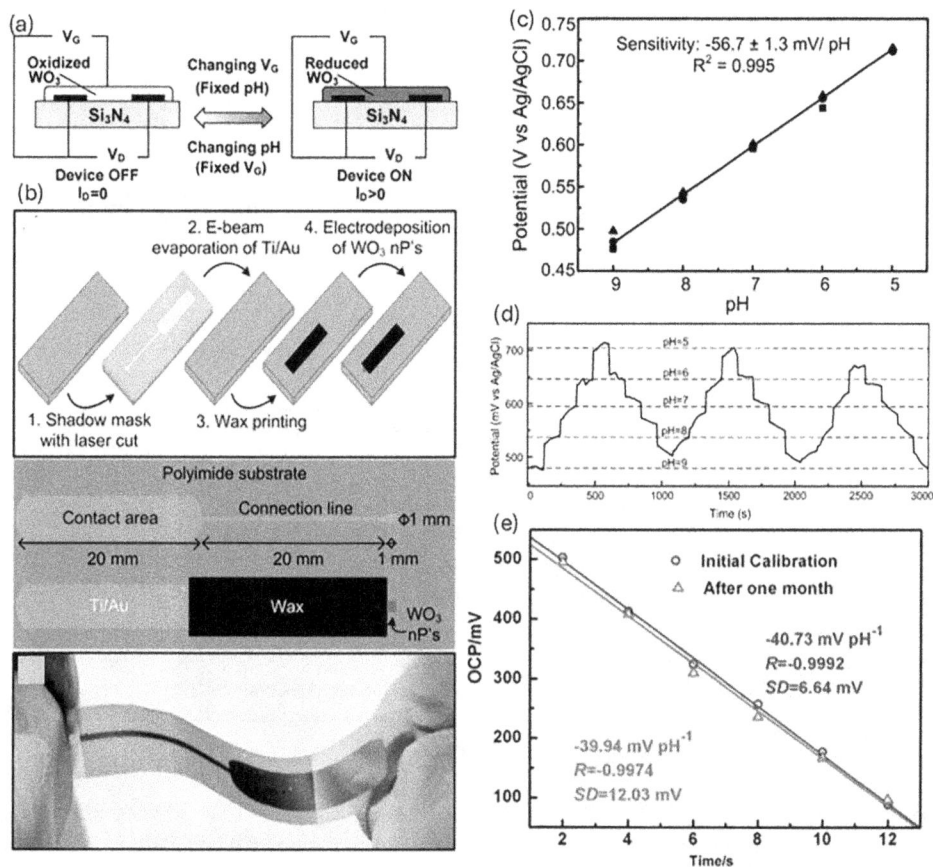

Figure 5.5. (a) Schematic of an electrochemical transistor function using WO_3 as a sensing layer. (b) Fabrication step, dimension of electrode and image of fabricated electrode of WO_3 flexible pH sensing layer, (c) Potentiometric performances of flexible sensor, (d) reversibility of performance of the flexible sensor, reprinted with permission from [61], copyright (2014) American Chemical Society, (e) potentiometric performances of WO_3/MWNTs electrode-based sensor, [29] John Wiley & Sons. copyright 2021 Wiley-VCH GmbH.

Based on this equation, the WO_3-based pH sensor follows a Nernstian response. There are multiple works reported on WO_3-based pH sensors, including the fabrication flexible sensor [2]. The steps of fabrication, the dimensions of electrodes and the image of a flexible pH-sensitive electrode, where the WO_3 layer was prepared by electrodeposition, are shown in figure 5.5(b) [61]. The developed sensor shows pH sensitivity of 56.7 mV/pH and excellent repeatability, as shown in figures 5.5(c) and (d) [61]. This flexible sensor is useful for wearable applications, especially physiological fluid monitoring. In one of the works, to improve reproducibility in performances, anti-interference properties and miniaturised fabrication, the WO_3 is incorporated with multiwalled carbon nanotube (MWCNT) [29]. The fabricated sensor shows a sub-Nernstian response with sensitivity of 41 mV/pH and it shows similar values even after one month stored in air, as shown in figure 5.5(e)

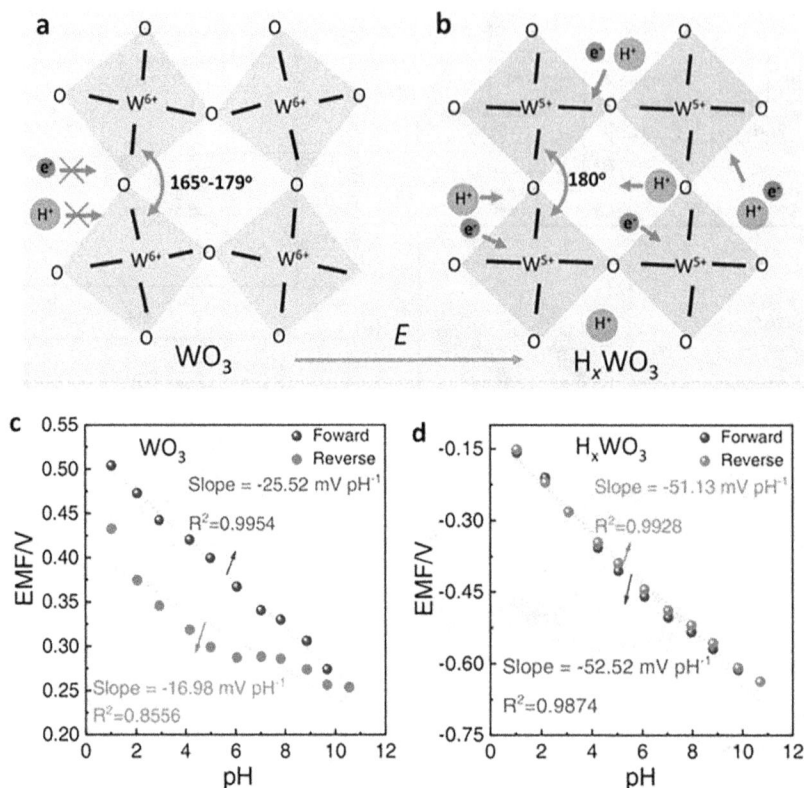

Figure 5.6. (a) Lattice proton intercalation of WO_3. (b) Structures of WO_3 before and after proton intercalation (H_xWO_3). (c) and (d) Linear response curves of the cyclic tests of WO_3 and H_xWO_3-based pH sensor [66], John Wiley & Sons. Copyright 2021 Wiley-VCH GmbH.

[29]. In an a-WO_3 as a gate ISFET based sensor, it was observed that the pH sensitivity increases with increasing temperature from 25 °C to 65 °C [65]. The drift rate increasing with increasing pH value and acid-base hysteresis width is smaller than basic-acid hysteresis. For this ISFET the hysteresis width depends on measuring loop time and measuring path [65].

Nowadays, there is a potential demand for microelectrochemical pH sensors, especially a miniaturised and reliable pH sensor for healthcare applications. Non-toxic materials like WO_3 have a key sensing role in such biosensors. In one of the recent works, the sensing of WO_3-based pH sensing layer was improved by one-step electrochemical proton intercalation, as shown figure 5.6(a) [66]. The proton intercalation enables changing monoclinic phase to the cubic phase of the WO_3, and it improves the proton exchange. This proton intercalation also supports improving the reversible response, fast response time, and good anti-interference. The proton intercalated H_xWO_3 shows a sensitivity of 51.13 mV/pH as compared to 25.52 mV/pH of the WO_3-based pH sensor, as shown in figures 5.6(b) and 5.6(c) [66]. The studies also reveal that the hysteresis deviation of H_xWO_3-based sensor (6.7 mV) is significantly lower than the WO_3-based pH sensor, which is 71.99 mV. In a

H_xWO_3-based sensitive electrode, the repeatability, response time and stability are much better than the WO_3-based pH-sensitive electrode [66]. This sensor shows excellent application in sweat monitoring and is discussed in chapter 9.

5.2.4 PbO₂-based pH sensor

For a reliable measurement in different environmental conditions, the highly electrocatalytically active and electrically conductive PbO_2 exhibits excellent opportunities. For electrochemical biosensing, including pH and creatinine monitoring, the PbO_2 as a sensitive electrode shows high sensitivity and selectivity, chemical stability and ease of fabrication [67–69]. The deficiency of oxygen or excess of lead in PbO_2 results in the characteristic metallic conductivity of lead oxide [2].

The internal redox reaction of PbO_2 electrode can be represented as [69, 70]

$$PbO_2 + H^+ + \bar{e} \Leftrightarrow PbOOH \qquad (5.14)$$

Hence the Nernst equation can be written as

$$E = E^0 + \frac{RT}{nF} \ln \frac{aPbO_2 aH^+}{aPbOOH} \qquad (5.15)$$

For a constant value of ($aPbO_2$) and ($aPbOOH$), and by substituting the values of R, T at 25 °C and F, the equation can be perfectly followed by Nernstian behaviour. Eftekhari [69] reported the possibility of the fabrication of pH-sensitive electrodes using the two crystalline structures of PbO_2, such as α-PbO_2 (more compact structure) and β-PbO_2 (more porous structure), on an aluminium substrate. Both electrode shows almost similar sensitivity and are close to 57 mV/pH. The developed pH sensor is used for the measurement of pH during flow Injection analysis (FIA). They carried out the sensor application in measuring pH values of soft drinks and fruit juices [69]. When the same crystalline structures of PbO_2 are coated on the top of a carbon ceramic electrode, the electrochemical properties are improved [70]. The electrochemical properties of the PbO_2 film improved at the carbon composite electrode. In the developed sensor, α-PbO_2 shows a sensitivity of 64.82 mV/pH, which is higher than the sensitivity value of 57.85 mV/pH for the β-PbO_2-based sensor [70]. The PbO_2 film coated on the carbon electrode-based sensor shows excellent response time, high selectivity, good stability and long-term usage [70

5.2.5 Lithium lanthanum titanate (LLTO)

For an industrial application including milk fermentation, yoghurt fabrication and waste water treatment, one of the best sensitive materials reported for pH-sensitive electrode fabrication is lithium lanthanum titanate (LLTO). The chemical stability, performance in high temperature and pressure and mechanical resistance endow ceramic LLTO as an excellent pH-sensitive material. In a preliminary work, Bohnke *et al* followed two methods for the preparation of ion-selective electrodes and electrochemical cells: membrane configuration, in which the internal reference was liquid and solid-state configuration, in which the internal reference was solid. It was found that, for the measurement temperatures of 20 °C and 60 °C, LLTO as a

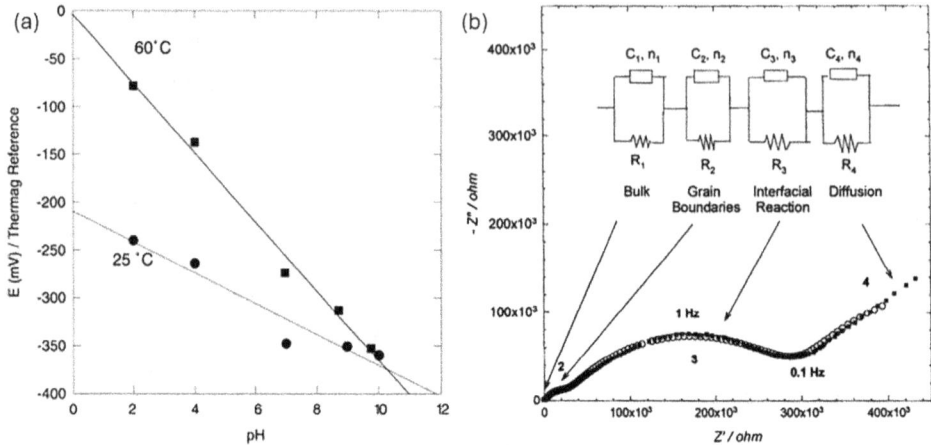

Figure 5.7. (a) Potentiometric pH sensing performances of LLTO at 25 °C and 60 °C, reprinted from [71], copyright (2003), with permission from Elsevier. (b) Electrochemical impedance spectroscopic analysis of LLTO with pH 2 reaction, reprinted from [72], copyright (2003), with permission from Elsevier.

solid-state configuration, the sensor shows sensitivity of 16 and 36 mV/pH (shown in figure 5.7(a)) and the sensor exhibits excellent performances in milk fermentation [71]. This ceramic material's electrode–electrolyte interaction was studied in the continued work of the authors through impedance spectroscopic analysis [72]. They observed that a complex reaction occurred during the LLTO-pH solution reaction and is shown in figure 5.7(b) [72]. They noticed that the grain boundaries of the electrode play a crucial role in the pH sensing [72]. The later studies of these materials show that the solid-state reaction small grain size of the powder, is favourable to the pH variation sensitivity and the sensitivity increases if the LLTO powder is ground before heat treatment. In addition to this, the sol–gel methods of preparation of LLTO are comparatively better than the solid-state reaction method in terms of producing a larger amount of pure LLTO [2].

5.2.6 Yitria stabilised zirconia (YSZ)

Compared with pH measurements at low temperature, the measurement at high temperature and pressure requires special attention in materials and the design. The pH of the solution needs to be monitored in geothermal applications, fossil power, nuclear power, oil well drilling scale deposition, crystal growth, corrosion and also in aqueous processing and hydrometallurgy [73, 74]. One of the promising materials implemented for such high-temperature and -pressure applications is yttria stabilised zirconia (YSZ) [73, 74]. The electrochemical measurement setup is highly important in such high-temperature measurements. Figure 5.8(a) shows one of the cells used for electrode testing and is based on the reported work [74]. The configuration employed for the YSZ-based pH sensor is shown in figure 5.8(a) [74].

The general theory of the YSZ-based pH sensor can be represented as [73, 74]

$$RE|H_2O, H^+|ZrO_2(Y_2O_3)|HgO|Hg0) \qquad (5.16)$$

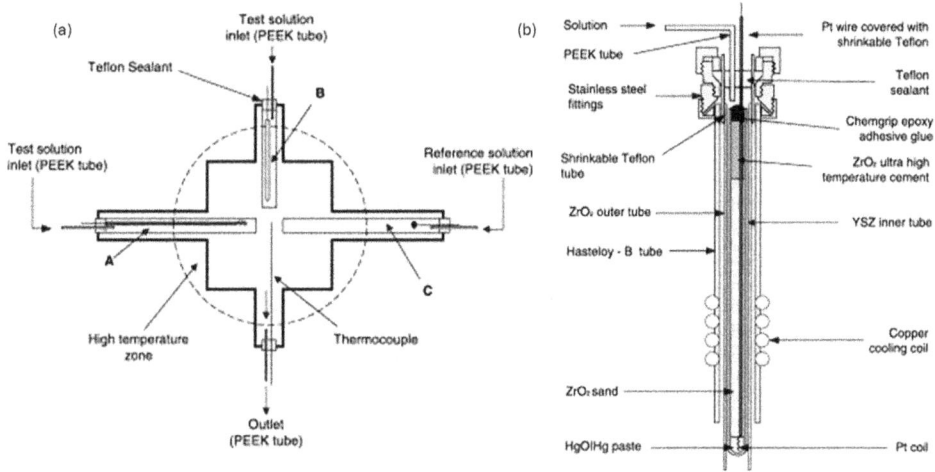

Figure 5.8. (a) The electrochemical cell setup for high-temperature measurement (b) design of flow-through YSZ pH sensor. Reprinted from [74], copyright (2003), with permission from Elsevier.

Here, RE is the reference electrode. The electrochemical reaction occurs at the external and internal membrane interface of the YSZ pH sensor, and it may be represented as [73, 74]

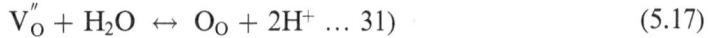

$$V_O^{''} + H_2O \leftrightarrow O_O + 2H^+ \dots 31) \qquad (5.17)$$

$$V_O^{''} + HgO + 2e^- \leftrightarrow O_O + Hg(32) \qquad (5.18)$$

Here $V_O^{''}$ denotes the oxygen ion vacancy on the surface of the ceramic and O_O an oxygen ion in a normal surface anion site. By the Nernst equation, the potential of the cell is given by

$$E_{YSZ} = E_{Hg/HgO}^O - \frac{2.303RT}{F}pH - \frac{2.303RT}{2F} \log H_2O + E_{RE}. \qquad (5.19)$$

The electromotive force of YSZ explains that when the membrane is in thermodynamic equilibrium, the cell potential is independent of the membrane properties [73, 74]. So far, several researchers have investigated the uses and applications of YSZ membrane at high temperature and high pressure. In one of the studies, it was found that the YSZ-based pH sensor enables measuring the pH of solutions up to 300 °C, and it obeys Nernstian behaviour [73]. In another study, there was noticed a flow-through electrochemical cell with a flow-through YSZ-based sensor and in which a flow-through external reference electrode was used that can measure pH at 250 °C in process sulfate systems and which is relevant to the hydrometallurgical process of nickeliferous laterites [75].

5.2.7 Other metal oxides

The materials based on Pt-group metals are very expensive, so the research interest has been directed towards the development of pH sensors by cheap materials, in

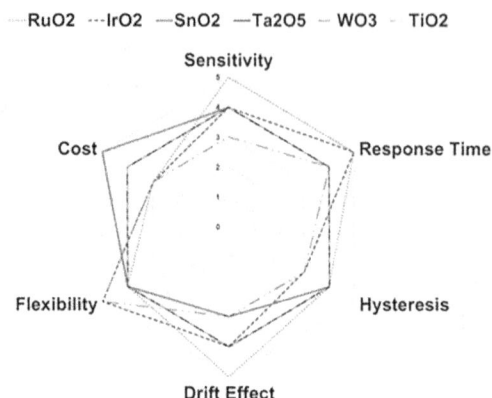

Figure 5.9. Comparison of pH sensing performances of various metal oxides. Reproduced from [2] CC BY 4.0.

which various materials, including Ta_2O_5, CuO, SnO_2, TiO_2, CeO_2 and composites are implemented for pH sensor fabrication. Some materials were reported in the previous reviews, and the performances of these MOx are given in table 5.2. The breakable glass pH electrode has limitations in its use for food processing, wearables and pollution-monitoring applications. A 'non-glass' pH sensor has greater demand in the food industry, biotechnology, water purification, etc, especially for online monitoring. So, these materials have a significant impact on the pH sensor designs due to their performance and electrochemical properties. Some of the materials are limited due to the hysteresis and drift effect of the sensing performance. To overcome this, composite structures also showed excellent properties. A comparison of the sensing performance of widely used MOx-based pH sensors is shown in figure 5.9. Comparison of the average sensitivities of various materials predicts that RuO_2-based sensors have very good sensing performance [2].

5.3 Summary

In electrochemical pH sensing, ion-sensitive metal oxides (MOx) have received substantial attention as sensitive electrodes, attributable to their elevated sensitivity, selectivity, stability, and prompt response. This chapter explores the significance and recent advancements associated with metal-metal oxide-based materials within the realm of electrochemical pH sensing.

References

[1] Głab S, Hulanicki A, Edwall G and Ingman F 1989 Metal–metal oxide and metal oxide electrodes as pH sensors *Crit. Rev. Anal. Chem.* **21** 29–47

[2] Manjakkal L, Szwagierczak D and Dahiya R 2020 Metal oxides based electrochemical pH sensors: current progress and future perspectives *Prog. Mater Sci.* **109** 100635

[3] Kurzweil P 2009 Metal oxides and ion-exchanging surfaces as pH sensors in liquids: state-of-the-art and outlook *Sensors* **9** 4955–85

[4] Song I, Fink K and Payer J 1998 Metal oxide/metal pH sensor: effect of anions on pH measurements *Corrosion* **54** 13–9

[5] Manjakkal L, Cvejin K, Kulawik J, Zaraska K, Szwagierczak D and Socha R P 2014 Fabrication of thick film sensitive RuO_2–TiO_2 and Ag/AgCl/KCl reference electrodes and their application for pH measurements *Sens. Actuators* B **204** 57–67

[6] Manjakkal L, Cvejin K, Kulawik J, Zaraska K, Szwagierczak D and Stojanovic G 2015 Sensing mechanism of RuO_2–SnO_2 thick film pH sensors studied by potentiometric method and electrochemical impedance spectroscopy *J. Electroanal. Chem.* **759** 82–90

[7] Fog A and Buck R P 1984 Electronic semiconducting oxides as pH sensors *Sens. Actuators* **5** 137–46

[8] Wang M, Yao S and Madou M 2002 A long-term stable iridium oxide pH electrode *Sens. Actuators* B **81** 313–5

[9] Huang W-D, Cao H, Deb S, Chiao M and Chiao J-C 2011 A flexible pH sensor based on the iridium oxide sensing film *Sens. Actuators* A **169** 1–11

[10] Marzouk S A, Ufer S, Buck R P, Johnson T A, Dunlap L A and Cascio W E 1998 Electrodeposited iridium oxide pH electrode for measurement of extracellular myocardial acidosis during acute ischemia *Anal. Chem.* **70** 5054–61

[11] Prats-Alfonso E, Abad L, Casañ-Pastor N, Gonzalo-Ruiz J and Baldrich E 2013 Iridium oxide pH sensor for biomedical applications. Case urea–urease in real urine samples *Biosens. Bioelectron.* **39** 163–9

[12] Martínez-Máñez R, Soto J, García-Breijo E, Gil L, Ibáñez J and Gadea E 2005 A multisensor in thick-film technology for water quality control *Sens. Actuators* A **120** 589–95

[13] Liao Y-H and Chou J-C 2008 Preparation and characteristics of ruthenium dioxide for pH array sensors with real-time measurement system *Sens. Actuators* B **128** 603–12

[14] Zhuiykov S 2009 Morphology of Pt-doped nanofabricated RuO_2 sensing electrodes and their properties in water quality monitoring sensors *Sens. Actuators* B **136** 248–56

[15] Manjakkal L, Cvejin K, Kulawik J, Zaraska K and Szwagierczak D 2014 The effect of sheet resistivity and storage conditions on sensitivity of RuO_2-based pH sensors *Key Eng. Mater., Trans Tech Publ.* 457–60

[16] Liao Y-H and Chou J-C 2009 Preparation and characterization of the titanium dioxide thin films used for pH electrode and procaine drug sensor by sol–gel method *Mater. Chem. Phys.* **114** 542–8

[17] Shin P-K 2003 The pH-sensing and light-induced drift properties of titanium dioxide thin films deposited by MOCVD *Appl. Surf. Sci.* **214** 214–21

[18] Zhao R, Xu M, Wang J and Chen G 2010 A pH sensor based on the TiO_2 nanotube array modified Ti electrode *Electrochim. Acta* **55** 5647–51

[19] Chou J C and Liao L P 2004 Study of TiO_2 thin films for ion sensitive field effect transistor application with RF sputtering deposition *Jpn. J. Appl. Phys.* **43** 61

[20] Tsai C N, Chou J C, Sun T P and Hsiung S K 2006 Study on the time-dependent slow response of the tin oxide pH electrode *IEEE Sens. J.* **6** 1243–9

[21] Chen P-Y, Yin L-T, Shi M-D and Lee Y-C 2013 Drift and light characteristics of EGFET based on SnO_2/ITO sensing gate *Life Sci. J.* **10** 3132–6

[22] Chin Y-L, Chou J-C, Sun T-P, Liao H-K, Chung W-Y and Hsiung S-K 2001 A novel SnO_2/Al discrete gate ISFET pH sensor with CMOS standard process *Sens. Actuators* B **75** 36–42

[23] Li H-H, Dai W-S, Chou J-C and Cheng H-C 2012 An extended-gate field-effect transistor with low-temperature hydrothermally synthesized nanorods as pH sensor *IEEE Electron Device Lett.* **33** 1495–7

[24] Manjakkal L, Cvejin K, Bajac B, Kulawik J, Zaraska K and Szwagierczak D 2015 Microstructural, impedance spectroscopic and potentiometric analysis of Ta_2O_5 electrochemical thick film pH sensors *Electroanalysis* **27** 770–81

[25] Chen M, Jin Y, Qu X, Jin Q and Zhao J 2014 Electrochemical impedance spectroscopy study of Ta_2O_5 based EIOS pH sensors in acid environment *Sens. Actuators* B **192** 399–405

[26] Kwon D-H, Cho B-W, Kim C-S and Sohn B-K 1996 Effects of heat treatment on Ta_2O_5 sensing membrane for low drift and high sensitivity pH-ISFET *Sens. Actuators* B **34** 441–5

[27] Chou J C and Wang Y F 2002 Preparation and study on the drift and hysteresis properties of the tin oxide gate ISFET by the sol–gel method *Sens. Actuators* B **86** 58–62

[28] Santos L, Neto J P, Crespo A, Nunes D, Costa N, Fonseca I M, Barquinha P, Pereira L, Silva J and Martins R 2014 WO_3 nanoparticle-based conformable pH sensor *ACS Appl. Mater. Interfaces* **6** 12226–34

[29] Zhang W-D and Xu B 2009 A solid-state pH sensor based on WO_3-modified vertically aligned multiwalled carbon nanotubes *Electrochem. Commun.* **11** 1038–41

[30] Yamamoto K, Shi G, Zhou T, Xu F, Zhu M, Liu M, Kato T, Jin J-Y and Jin L 2003 Solid-state pH ultramicrosensor based on a tungstic oxide film fabricated on a tungsten nano-electrode and its application to the study of endothelial cells *Anal. Chim. Acta* **480** 109–17

[31] Chou J-C and Chiang J-L 2000 Ion sensitive field effect transistor with amorphous tungsten trioxide gate for pH sensing *Sens. Actuators* B **62** 81–7

[32] Lale A, Tsopela A, Civélas A, Salvagnac L, Launay J and Temple-Boyer P 2015 Integration of tungsten layers for the mass fabrication of WO_3-based pH-sensitive potentiometric microsensors *Sens. Actuators* B **206** 152–8

[33] Maiolo L, Mirabella S, Maita F, Alberti A, Minotti A, Strano V, Pecora A, Shacham-Diamand Y and Fortunato G 2014 Flexible pH sensors based on polysilicon thin film transistors and ZnO nanowalls *Appl. Phys. Lett.* **105** 093501

[34] Fulati A, Usman Ali S M, Riaz M, Amin G, Nur O and Willander M 2009 Miniaturized pH sensors based on zinc oxide nanotubes/nanorods *Sensors* **9** 8911–23

[35] Al-Hilli S M, Willander M, Öst A and Stralfors P 2007 ZnO nanorods as an intracellular sensor for pH measurements *J. Appl. Phys.* **102** 4304

[36] Kao C H, Chen H, Lee M L, Liu C C, Ueng H-Y, Chu Y C, Chen Y J and Chang K M 2014 Multianalyte biosensor based on pH-sensitive ZnO electrolyte–insulator–semiconductor structures *J. Appl. Phys.* **115** 184701

[37] Li H-H, Yang C-E, Kei C-C, Su C-Y, Dai W-S, Tseng J-K, Yang P-Y, Chou J-C and Cheng H-C 2013 Coaxial-structured ZnO/silicon nanowires extended-gate field-effect transistor as pH sensor *Thin Solid Films* **529** 173–6

[38] Korotcov A V, Huang Y S, Tiong K K and Tsai D S 2007 Raman scattering characterization of well-aligned RuO_2 and IrO_2 nanocrystals *J. Raman Spectrosc.* **38** 737–49

[39] Meng L-j, Teixeira V and Dos Santos M 2003 Raman spectroscopy analysis of magnetron sputtered RuO_2 thin films *Thin Solid Films* **442** 93–7

[40] Singh K, Chen C-H, Tai L-C, Pang S-T and Pan T-M 2024 Enhancing pH sensing capabilities through hydroxylated surface groups on RuOx flexible EGFET sensor *IEEE Sens. J.* **24** 13863–9

[41] Shylendra S P, Wajrak M, Alameh K and Kang J J 2024 Effect of post deposition annealing on the sensing properties of thin film ruthenium oxide (RuO_2) pH sensor *IEEE Sens. J.* **24** 12498–503

[42] McMurray H N, Douglas P and Abbot D 1995 Novel thick-film pH sensors based on ruthenium dioxide-glass composites *Sens. Actuators* B **28** 9–15

[43] Trasatti S 1991 Physical electrochemistry of ceramic oxides *Electrochim. Acta* **36** 225–41

[44] Kurzweil P 2009 Precious metal oxides for electrochemical energy converters: pseudocapacitance and pH dependence of redox processes *J. Power Sources* **190** 189–200

[45] Mihell J and Atkinson J 1998 Planar thick-film pH electrodes based on ruthenium dioxide hydrate *Sens. Actuators* B **48** 505–11

[46] Zhuiykov S 2009 Morphology of Pt-doped nanofabricated RuO_2 sensing electrodes and their properties in water quality monitoring sensors *Sens. Actuators* B **136** 248–56

[47] Manjakkal L, Cvejin K, Kulawik J, Zaraska K and Szwagierczak D 2013 A low-cost ph sensor based on RuO_2 resistor material *Nano Hybrids* **5** 1–15

[48] Manjakkal L, Cvejin K, Kulawik J, Zaraska K and Szwagierczak D 2014 A comparative study of potentiometric and conductimetric thick film pH sensors made of RuO_2 pastes *Sens. Lett.* **12** 1645–50

[49] Manjakkal L, Cvejin K, Kulawik J, Zaraska K and Szwagierczak D 2014 The effect of sheet resistivity and storage conditions on sensitivity of RuO_2-based pH sensors *Key Eng. Mater.* **605** 457–60

[50] Manjakkal L, Zaraska K, Cvejin K, Kulawik J and Szwagierczak D 2016 Potentiometric RuO_2–Ta_2O_5 pH sensors fabricated using thick film and LTCC technologies *Talanta* **147** 233–40

[51] Pocrifka L A, Gonçalves C, Grossi P, Colpa P C and Pereira E C 2006 Development of RuO_2–TiO_2 (70–30)mol% for pH measurements *Sens. Actuators* B **113** 1012–6

[52] Manjakkal L, Synkiewicz B, Zaraska K, Cvejin K, Kulawik J and Szwagierczak D 2016 Development and characterization of miniaturized LTCC pH sensors with RuO_2-based sensing electrodes *Sens. Actuators* B **223** 641–9

[53] Liu X, Pei Y, Wang Y, Liu X, Chen X, Sun C and Liu N 2022 Performance of IrOx pH sensor prepared by electrochemical and thermal oxidation *IEEE Sens. J.* **22** 12560–9

[54] Katsube T, Lauks I and Zemel J N 1981 pH-sensitive sputtered iridium oxide films *Sens. Actuators* **2** 399–410

[55] Kreider K 1991 Iridium oxide thin-film stability in high-temperature corrosive solutions *Sens. Actuators* B **5** 165–9

[56] Korotcov A V, Huang Y-S, Tiong K-K and Tsai D-S 2007 Raman scattering characterization of well-aligned RuO_2 and IrO_2 nanocrystals *J. Raman Spectrosc.* **38** 737–49

[57] Cai Z 2005 Mechanism of electrochemical evolution of chlorine and oxygen at Ti/SnO_2–IrO_2 anode *Prog. Nat. Sci.* **15** 76–81

[58] Kuznetsova E, Petrykin V, Sunde S and Krtil P 2015 Selectivity of nanocrystalline IrO_2-based catalysts in parallel chlorine and oxygen evolution *Electrocatalysis* **6** 198–210

[59] Yin J, Chen W, Gao W, Zhang X, Tang B and Jin Q 2023 Batch Fabrication of microminiaturized pH sensor integrated with IrO_x film and solid state Ag/AgCl electrode for tap water quality online detection *IEEE Sens. J.* **23** 3475–84

[60] Kinlen P J, Heider J E and Hubbard D E 1994 A solid-state pH sensor based on a Nafion-coated iridium oxide indicator electrode and a polymer-based silver chloride reference electrode *Sens. Actuators* B **22** 13–25

[61] Santos L *et al* 2014 WO_3 nanoparticle-based conformable pH sensor *ACS Appl. Mater. Interfaces* **6** 12226–34

[62] Kuo C-Y, Wang S-J, Ko R-M and Tseng H-H 2018 Super-Nernstian pH sensors based on WO$_3$ nanosheets *Jpn. J. Appl. Phys.* **57** 04FM09

[63] Salazar P, Garcia-Garcia F J, Yubero F, Gil-Rostra J and González-Elipe A R 2016 Characterization and application of a new pH sensor based on magnetron sputtered porous WO$_3$ thin films deposited at oblique angles *Electrochim. Acta* **193** 24–31

[64] Natan M J, Mallouk T E and Wrighton M S 1987 The pH-sensitive tungsten(VI) oxide-based microelectrochemical transistors *J. Phys. Chem.* **91** 648–54

[65] Chiang J-L, Jan S-S, Chou J-C and Chen Y-C 2001 Study on the temperature effect, hysteresis and drift of pH-ISFET devices based on amorphous tungsten oxide *Sens. Actuators* B **76** 624–8

[66] Tang Y *et al* 2022 Lattice proton intercalation to regulate WO$_3$-based solid-contact wearable pH sensor for sweat analysis *Adv. Funct. Mater.* **32** 2107653

[67] Jeon S M, Lee H B, Ha C H, Kim D H, Li C A, Song S H, Lee C-J, Han D K and Seong G H 2024 High-performance electrochemical creatinine sensors based on β-lead dioxide/single-walled carbon nanotube electrodes *Anal. Chem.* **96** 15941–9

[68] Lima A C, Jesus A A, Tenan M A, Silva A F d S and Oliveira A F 2005 Evaluation of a high sensitivity PbO$_2$ pH-sensor *Talanta* **66** 225–8

[69] Eftekhari A 2003 pH sensor based on deposited film of lead oxide on aluminum substrate electrode *Sens. Actuators* B **88** 234–8

[70] Razmi H, Heidari H and Habibi E 2008 pH-sensing properties of PbO$_2$ thin film electro-deposited on carbon ceramic electrode *J. Solid State Electrochem.* **12** 1579–87

[71] Bohnke C, Duroy H and Fourquet J L 2003 pH sensors with lithium lanthanum titanate sensitive material: applications in food industry *Sens. Actuators* B **89** 240–7

[72] Bohnke C and Fourquet J L 2003 Impedance spectroscopy on pH-sensors with lithium lanthanum titanate sensitive material *Electrochim. Acta* **48** 1869–78

[73] Macdonald D D, Hettiarachchi S and Lenhart S J 1988 The thermodynamic viability of yttria-stabilized zirconia pH sensors for high temperature aqueous solutions *J. Solution Chem.* **17** 719–32

[74] Lvov S N, Zhou X Y, Ulmer G C, Barnes H L, Macdonald D D, Ulyanov S M, Benning L G, Grandstaff D E, Manna M and Vicenzi E 2003 Progress on yttria-stabilized zirconia sensors for hydrothermal pH measurements *Chem. Geol.* **198** 141–62

[75] Seneviratne D S, Papangelakis V G, Zhou X Y and Lvov S N 2003 Potentiometric pH measurements in acidic sulfate solutions at 250 °C relevant to pressure leaching *Hydrometallurgy* **68** 131–9

[76] Xu B and Zhang W-D 2010 Modification of vertically aligned carbon nanotubes with RuO$_2$ for a solid-state pH sensor *Electrochim. Acta* **55** 2859–64

[77] Koncki R and Mascini M 1997 Screen-printed ruthenium dioxide electrodes for pH measurements *Anal. Chim. Acta* **351** 143–9

[78] Pocrifka L, Gonçalves C, Grossi P, Colpa P and Pereira E 2006 Development of RuO$_2$–TiO$_2$ (70–30) mol% for pH measurements *Sens. Actuators* B **113** 1012–6

[79] Karimi Shervedani R, Zare Mehrdjardi H and Kazemi Ghahfarokhi S 2007 Electrochemical characterization and application of Ni–RuO$_2$ as a pH sensor for determination of petroleum oil acid number *J. Iran. Chem. Soc.* **4** 221–8

[80] Zhuiykov S, Kats E, Kalantar-zadeh K, Breedon M and Miura N 2012 Influence of thickness of sub-micron Cu$_2$O-doped RuO$_2$ electrode on sensing performance of planar electrochemical pH sensors *Mater. Lett.* **75** 165–8

[81] Zhuiykov S 2010 Development of ceramic electrochemical sensor based on $Bi_2Ru_2O_{7+x}^-$ RuO_2 sub-micron oxide sensing electrode for water quality monitoring *Ceram. Int.* **36** 2407–13

[82] Labrador R H, Soto J, Martínez-Máñez R, Coll C, Benito A, Ibáñez J, García-Breijo E and Gil L 2007 An electrochemical characterization of thick-film electrodes based on RuO_2-containing resistive pastes *J. Electroanal. Chem.* **611** 175–80

[83] Atkinson J K, Cranny A, Glasspool W V and Mihell J A 1999 An investigation of the performance characteristics and operational lifetimes of multi-element thick film sensor arrays used in the determination of water quality parameters *Sens. Actuators* B **54** 215–31

[84] Da Silva G, Lemos S, Pocrifka L, Marreto P, Rosario A and Pereira E 2008 Development of low-cost metal oxide pH electrodes based on the polymeric precursor method *Anal. Chim. Acta* **616** 36–41

[85] Pan C-W, Chou J-C, Sun T-P and Hsiung S-K 2005 Development of the tin oxide pH electrode by the sputtering method *Sens. Actuators* B **108** 863–9

[86] Schöning M J, Brinkmann D, Rolka D, Demuth C and Poghossian A 2005 CIP (cleaning-in-place) suitable 'non-glass' pH sensor based on a Ta_2O_5-gate EIS structure *Sens. Actuators* B **111** 423–9

[87] Chen M, Jin Y, Qu X, Jin Q and Zhao J 2014 Electrochemical impedance spectroscopy study of Ta_2O_5-based EIOS pH sensors in acid environment *Sens. Actuators* B **192** 399–405

[88] Yoshida S, Hara N and Sugimoto K 2004 Development of a wide range pH sensor based on electrolyte-insulator-semiconductor structure with corrosion-resistant Al_2O_3 Ta_2O_5 and Al_2O_3 ZrO_2 double-oxide thin films *J. Electrochem. Soc.* **151** H53

[89] Betelu S, Polychronopoulou K, Rebholz C and Ignatiadis I 2011 Novel CeO_2-based screen-printed potentiometric electrodes for pH monitoring *Talanta* **87** 126–35

[90] Qingwen L, Guoan L and Youqin S 2000 Response of nanosized cobalt oxide electrodes as pH sensors *Anal. Chim. Acta* **409** 137–42

[91] Vijayakumar M, Pham Q N and Bohnke C 2005 Lithium lanthanum titanate ceramic as sensitive material for pH sensor: influence of synthesis methods and powder grains size *J. Eur. Ceram. Soc.* **25** 2973–6

[92] Zhao R, Xu M, Wang J and Chen G 2010 A pH sensor based on the TiO_2 nanotube array modified Ti electrode *Electrochim. Acta* **55** 5647–51

[93] Chen Y, Mun S C and Kim J 2013 A wide range conductometric pH sensor made with titanium dioxide/multiwall carbon nanotube/cellulose hybrid nanocomposite *IEEE Sens. J.* **13** 4157–62

[94] Manjakkal L, Djurdjic E, Cvejin K, Kulawik J, Zaraska K and Szwagierczak D 2015 Electrochemical Impedance Spectroscopic Analysis of RuO_2 Based Thick Film pH Sensors *Electrochim. Acta* **168** 246–255

IOP Publishing

Advanced Electrochemical pH Sensing Technologies
Scientific fundamentals and applications
Libu Manjakkal

Chapter 6

Carbon-based pH sensors: sensor fabrication and performance

6.1 Introduction

In electrochemical pH sensor fabrication, major attention has been focussed on the development of new sensing electrodes using nanomaterials due to their large number of binding sites and surface area for selective sensing. The signal amplification for specific chemical analytes is important in sensor development. Even though the MOx-based materials have been found to have excellent sensitivity, selectivity and stability for electrochemical pH sensing, for a disposable, flexible, and low-cost fabrication, carbon-based electrodes have been found to be an excellent alternative sensitive material. Carbon-based electrodes offer high electrical conductivity, high specific surface area, low weight and small sizes, high porosity, excellent chemical and thermal stability and low cost and mass production possibilities [1–3]. Carbon-based electrodes can also be used in composite electrodes to enhance the conductivity and increase the electrochemical double-layer capacitance (EDLC) of the electrodes [1]. Due to the high electron transportation, excellent mechanical stability and conductivity, carbon-based materials have been implemented for various bio-sensors and electrochemical sensors fabrication [4–6]. The quick electron transportation of the carbon-based electrodes enables a high signal-to-noise ratio in electrochemical and biosensors. However, the stability of the carbon-based sensitive electrode in repeated use reduces its practical implementation [7]. These electrodes will be excellent, sensitive electrodes for disposable biomedical applications.

Various allotropes of carbon nanomaterials, including activated carbon, graphene, carbon nanotubes (CNTs), diamond-like carbon, carbon nanodots, carbon nanofibres and carbon quantum dots (CQDs) are found to have exceptional electrochemical properties for sensing [8–10]. The electron transfer rates of the carbon electrode between the electrodes and electrolyte (e.g. $Fe[(CN)_6]^{4-/3-}$) depend on the different types of carbon allotropes, as shown in figure 6.1 [11]. Many manufacturing methods

doi:10.1088/978-0-7503-6079-1ch6

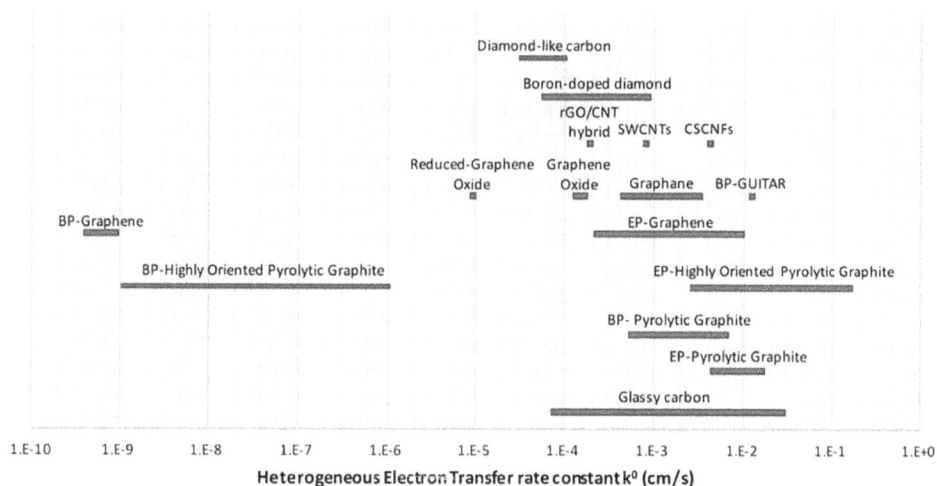

Figure 6.1. Heterogeneous electron transfer rates for different carbon allotropes in the $Fe[(CN)_6]^{4-/3-}$ benchmark redox pair sensitive to surface structures in the carbonaceous electrodes (BP—Basal plane and EP—Edge-oriented plane). Reprinted from [11], copyright (2017), with permission from Elsevier.

are implemented for the preparation of various carbon structures [12]. These carbon allotropes are reported to have generally 3D, 2D, 1D and 0D dimensions and are synthesised based on top-down, bottom-up, solid state, solution process, vapourisation, wet chemical methods, etc [13, 14]. The major synthesis approach of carbon allotropes in nanoscale structures is given in the reported work [13].

For electrochemical sensor fabrication, various allotropes of carbon are used [11, 14, 15], and a few of them are schematically represented in figure 6.2. Among these allotropes CNTs, carbon-based pastes, graphene oxides etc, are used for the fabrication of electrochemical pH sensors. This chapter focuses on the discussion of various carbon structures for pH sensor development and performance. Carbon is used as a sensing, counter and reference electrode material in the pH sensor fabrication. This chapter provides an overview of different types of pH sensors developed using carbon as sensitive electrode and also provides information about carbon-based composites.

The selection of sensing, counter and reference electrode materials is the most important factor for the fabrication of an excellent electrochemical pH sensor. Major attention has been focused on the development of electrochemical sensing using nanomaterials due to their large number of binding sites and surface area for selective sensing. In sensor fabrication, the selectivity, high sensitivity, limit of detection and low cost are other important factors.

6.2 Carbon nanotube-based pH sensor

CNTs are generally graphene sheets rolled into cylinders with a diameter of approximately one nanometre. CNTs are mainly single-walled CNTs (SWCNTs) and multiwalled CNTs (MWCNTs), in which SWCNTs typically have a diameter of

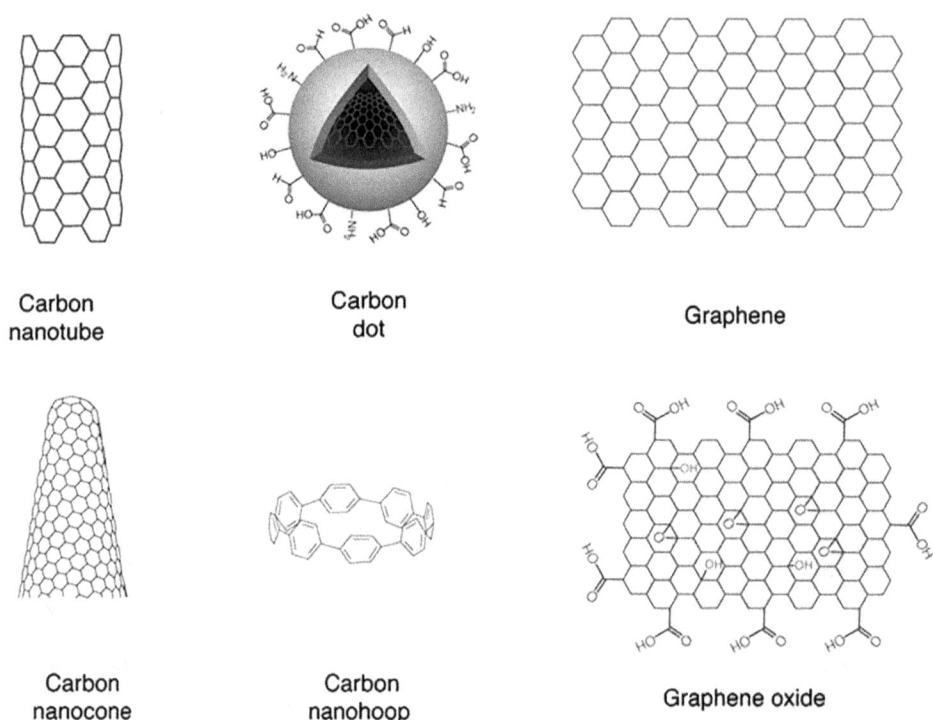

| Carbon nanotube | Carbon dot | Graphene |

| Carbon nanocone | Carbon nanohoop | Graphene oxide |

Figure 6.2. Allotropes of carbon. Reproduced from [15] CC BY 4.0.

1.4 nm and MWCNTs are in the order of 10–20 nm diameter and interspacing of 3.4 Å [16]. CNTs act as either a metal or a semiconductor based on their atomic structure, diameter and helicity (chiral angle) [17]. Due to their excellent electronic, mechanical and chemical properties, CNTs are widely used now for electrochemical and biosensors. CNTs are suitable for the modification of electronic conductivity and electron transfer reaction of various electrodes, and also exhibit excellent electrochemical and chemical stabilities [18, 19]. These unique properties lead CNTs to application as excellent pH-sensitive electrodes, especially for flexible and disposable applications [20]. The variation of resistance with the pH value of the solution of CNTs coated on the top of the paper substrate works as an excellent chemiresistive pH sensor. The chemiresistive pH sensor in figure 6.3(a) shows that the resistance variation of the CNT bundle is proportional to the pH of the analyte solution. Here, when CNT interacts with the solution, a peak in the density of state arises at the Fermi level and the energy gap is significantly reduced due to the interaction between the oxygen and carbon atoms [20]. The OH group can form an acceptor level, and it increases the conductivity. Hence, with a higher pH value (moving towards basic) the conductivity enhances or the resistance decreases, as shown in figure 6.3(b) [20]. Such paper-based CNT-coated sensors could be useful for disposable bioassays in wearable applications.

Various methods and different designs were employed for CNT-based pH sensor fabrication. In one of the works, aerosol jet printing offers a serpentine-shaped

(a)

(b)

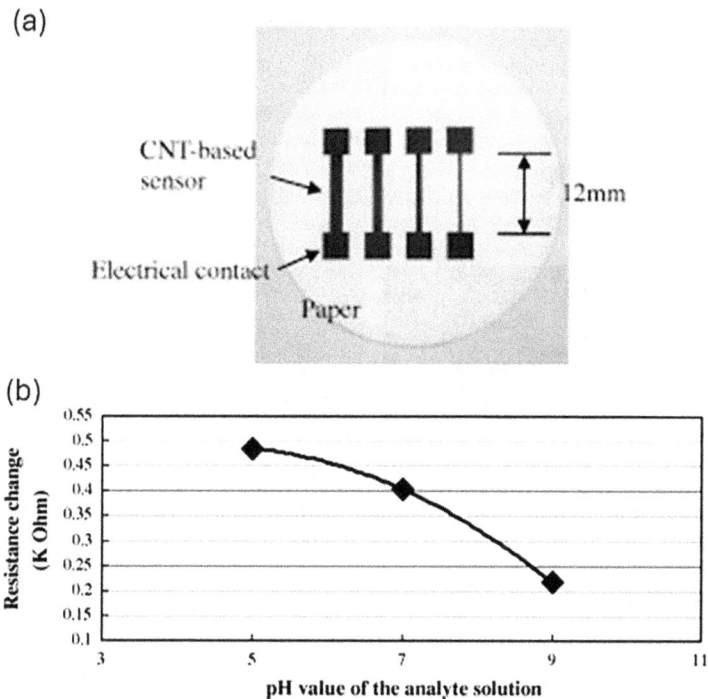

Figure 6.3. (a) Schematic representation of CNT-based pH sensor (b) the variation of resistance with pH value of CNT-based sensor. Reprinted from [20], copyright (2012), with permission from Elsevier.

chemiresistive pH sensor, which shows sensitivity of 59 kΩ/pH with a response time of 20 s [3]. The fine resolution printing of this approach will enable deposition of CNTs for high-performance sensing devices [3]. In a screen-printed flexible electrode, the surface functionalised MWCNT (f-MWCNT) was drop-cast as a pH-sensitive electrode. For cost-effective mass fabrication, spray-coated SWCNTs were used as a pH sensor. In this spray coating, the SWCNTs were either grown on a catalysed substrate or a solution of suspended SWCNTs was directly coated on the substrate. The random formation of CNT network reduces the reproducibility in sensing performances, and to overcome this issue dielectrophoresis (DEP) method was employed. DEP allows aligned CNT formation, positioning of CNTs into a precise location, which leads to obtaining a highly conductive CNT network. This DEP was employed for the development of a CNT-based (SWCNT or MWCNT) chemiresistive pH sensor [21, 22]. In one of the works for pH sensing, SWCNTs were surface-treated with nitric acid for surface carboxyl-functionalisation. The H^+ and OH^- from the pH buffer solution interact with the carboxyl group leading to holes and electrons generated in the SWCNT, which changes the resistance of the chemiresistive pH sensor [22]. The fabrication steps and the developed chemiresistive pH sensor are shown in figure 6.4(a) [22]. In the range of pH 5–9, this sensor shows a linear relationship of $R/R_0 = 0.285 \times$ pH $- 0.942$ as shown in figure 6.4(b) [22].

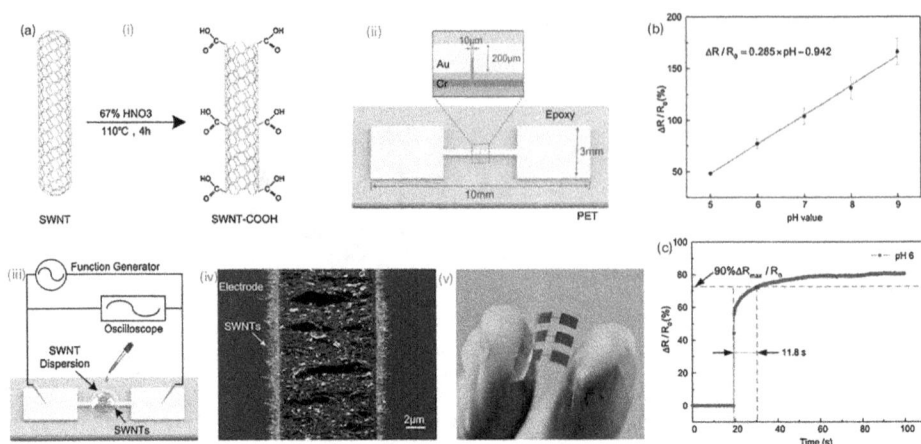

Figure 6.4. (a) Schematic of the process of the fabrication of SWCNT-based chemiresistive pH sensor by DEP method. (i) Surface functionalization of a SWCNT. (ii) Dimensions of the microelectrode for the sensor fabrication. (iii) Process of DEP assembly of SWCNTs onto an electrode gap. (iv) SEM image of DEP assembled SWCNT. (v) Image of the fabricated flexible chemiresistive pH sensor on the PET substrate. (b) The linear relationship between the sensor response and the pH values from pH 5 to 9. (c) The response time measurement under the pH-6 buffer solution. Reprinted from [22], copyright (2016), with permission from Elsevier.

It was found that the response time of the sensor varied in the range of 0.2–22.6 s with increasing pH value of the solution. For a pH of 6, the sensor shows a response time of 11.8 s, as shown in figure 6.4(c) [22].

6.3 Graphene-based pH sensor

Graphene is a single atomic plane of sp^2-bonded carbon atoms with a hexagonal configuration [23, 24]. Graphene exhibits a very large surface area 2630 m^2 g^{-1}, thermal conductivity ($\sim$5000 W m^{-1} K^{-1}), high Young's modulus ($\sim$1.0 TPa) and high mechanical strength [23, 24]. The large 2D electrical conductivity and surface area of the graphene lead to its applications in electrochemical and biosensors. As compared to CNT, graphene doesn't have metallic impurities [23]. In electrochemical and biosensing, the graphene or graphene-based composite is used as a sensitive electrode and is described in [23]. The high reactivity of graphene to oxygen-bearing groups like OH$^-$ and its high electrochemical performance have led them to be an excellent pH-sensitive electrode compared with CNTs, graphite and glass carbon-based electrodes. Graphene has a generally hydrophobic nature and hence its surface interaction with ions in the solution is extremely limited. To convert into a hydrophilic nature, defects are created in the surface of the graphene either through dopants or functionalisation [25]. The surface functionalisation is responsible for the pH sensing performance of the graphene. As an example, the carboxyl and amine groups are responsible for the response at low pH values, while the

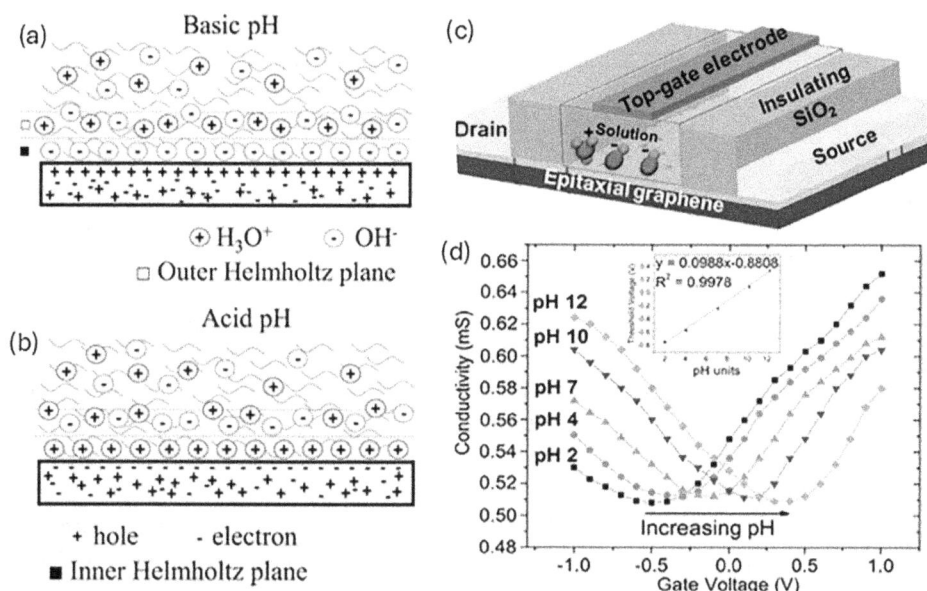

Figure 6.5. (a) and (b) Distribution of holes and electrons in the graphene-solution EDL in basic and acidic pH, reprinted from [24], copyright (2018), with permission from Elsevier. (c) Schematic representation of SGFET based on epitaxial graphene as channel layer, reprinted from [26], copyright (2018), with permission from Elsevier. (d) Sensing performances of SDFET fabricated with 3–4 layers of epitaxial graphene layers the conductivity versus gate potential plots, reprinted from [26], copyright (2008), with permission from Elsevier.

hydroxyl groups are responsible for the response at high pH values [25]. The hydronium and hydroxide ions in the solution interact with the surface of the graphene and lead to forming an electric double layer (EDL) [25]. If the inner Helmholtz layer is dominated by adsorbed H_3O^+, the graphene induces a negative charge and acts as an n-dopant [25]. However, if OH dominates the inner Helmholtz layer, the graphene induces positive charges and acts as a p-dopant [25]. The distribution of holes and electrons in the graphene-pH solution EDL in basic and acidic pH is shown in figures 6.5(a) and (b) [24].

Graphene is mainly used as a sensitive electrode in a field-effect transistor (FET)-based pH sensor and chemiresistive-based pH sensors. In an FET-based pH sensor, the graphene acts as a channel layer. In one of the preliminary works, epitaxial graphene (graphene grown directly on a crystalline substrate) works as a channel layer in a solution-gate FET (SGFET) [26], as shown in figure 6.5(c). In this graphene-based liquid gate FET, the variation in channel conductance is achieved by applying a gate potential from the reference electrode (top of the channel) and across the electrolyte. When testied in pH solution, both OH^- and H_3O^+ led to modulating the channel conductance by doping holes or electrons [26]. Here, the graphene/electrolyte interface allows the capacitive charging of the surface by H_3O^+ or OH^-. Such SGFET exhibits very high sensitivity—99 mV/pH as shown in figure 6.5(d) where the conductivity varies with pH value of the solution [26].

A common configuration of SGFET is shown in figure 6.5(c). Based on the SGFET, there are many works reported in electrolyte-gated FET for pH and other biological parameters monitoring [27]. These works were carried out with phosphate buffer solution (PBS) with a range from 20–50 mM [28]. It was found that the concentration of ions in the test PBS significantly influences the pH sensing performance [28]. In a chemical vapour deposition (CVD), synthesised graphene-based FET shows that the pH sensor effectively operates in high ionic concentration solutions with minimum sensitivity fluctuations [28]. A detailed study of an FET-based pH sensor is provided chapter 8.

The material preparation, the costly device fabrication and the integration of miniaturised reference electrode are the critical challenges in the graphene-FET-based pH sensor. The microfabrication approaches, such as lithography and etching, costly physical and chemical deposition methods, complex masking and aligning approaches, are the fabrication challenges. In addition, the graphene acts as a channel layer and is a semiconducting material in FET, and it needs a gate electrode. One of the simple approaches introduced in the pH sensor is a chemiresistive-based configuration [29]. One of the simple mask-free fabrication processes of graphene-based chem-resistive pH sensor is shown in figure 6.6(a) [29]. Here, the graphene sheets are made by the mechanical exfoliation method. When a pH solution interacteds with the graphene layer, a linear relationship between the pH value and the change in resistance value of the sensors was observed [29]. The sensor exhibited a sensitivity of 2.13 kΩ/pH in the range of pH value 4–9, as shown in figure 6.6(b) [29].

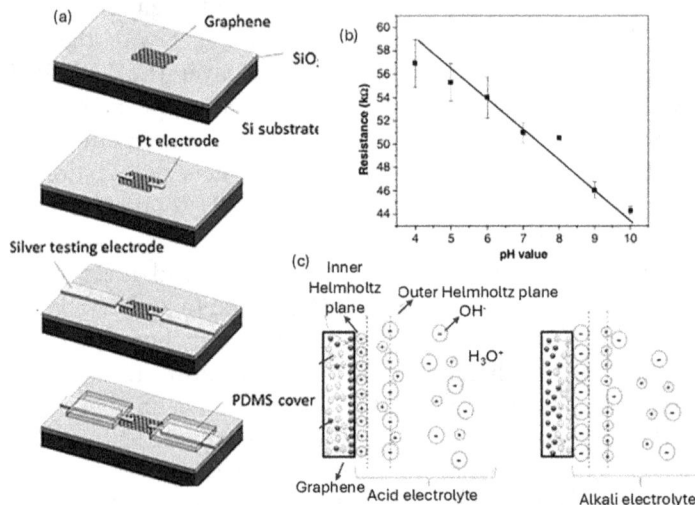

Figure 6.6. (a) Schematic representation of the fabrication of a graphene-based chemiresistive pH sensor. (b) Variation of resistance with pH value of the solution for the sensor. (c) H$_3$O$^+$ is attached to the inner Helmholtz plane in acidic solution, and OH$^-$ is attached to the inner Helmholtz plane in alkali electrolyte. Reproduced from [29], copyright IOP Publishing Ltd. All rights reserved.

As discussed above, the formation of EDL shown in figure 6.6(c) with the H_3O^+ leading to making graphene n-doped, and OH^- making graphene p-doped are responsible for the pH sensing mechanism [29]. In another chemiresistive pH sensor, a graphite powder ink prepared via sonication drop casting on the top of the electrode is designed as a sensor [25]. The team investigated the chemiresistive response of the graphene prepared in different sonication times [25]. For a multi-layer graphene prepared in 6, 12 and 18 h sonicated samples it is found that the sensitivity increases with an increase in sonication time. The 18 h sonicated sample shows high sensitivity and high response time [25].

6.4 Graphene oxide-based pH sensor

Graphene oxide (GO) is a layered carbon structure with oxygen-containing functional groups such as $=O$, $-OH$, $-O-$ and $-COOH$, which are attached to both sides of the layer as well as the edges of the plane, as shown in figure 6.2. This structure enables expansion of interlayer distance and makes the atomic-thick layers hydrophilic [30]. Due to the unique chemical structure, GO can easily be functionalised for electrochemical and biomedical applications [31]. The presence of functional groups in GO provides an electrochemical active surface area for the development of electrochemical devices

In pH sensing, the $-OH$ and $-COOH$ functional groups are responsible for sensing, and these groups can exchange hydrogen ions [32]

$$GO - O^-/GO - COO^- + H^+ + 1e^- = GO - OH/GO - COOH$$

The Nernst equation for this reversible reaction can be written as

$$E_{GO} = E_{constant} - 0.0592 \, pH$$

For a potentiometric pH sensor in one of the works, the GO solution is coated on the top of the graphitic carbon electrode, which acts as a pH-sensitive electrode and a saturated calomel electrode (SCE) as a reference electrode [32]. The fabricated sensor exhibits sensitivity of 51.1 mV/pH in the pH range of 2–10, as shown in figure 6.7(a) [32].

The monitoring of pH along with the temperature is highly important in the sensing mechanism for correlating its variation. In wearable devices, the integration of both pH and temperature sensors in a single substrate is ideal for predicting the human body fluid parameters. Such integrated sensors will be significant in real-time and continuous monitoring of body fluid parameters [2]. In one of the works, GO is employed as a pH-sensitive electrode, and reduced GO is used as temperature temperature-sensitive electrode in an integrated sensor on a flexible substrate, as shown in figure 6.7(b). In a GO-based pH sensor, the hydroxyl and carboxyl groups protonate or deprotonate during pH changes [2]. The developed pH sensor exhibits a pH sensitivity of 40 ± 4 mV/pH in a pH range of 4–10, as shown in figure 6.7(c), and temperature sensors reveal a sensitivity of $10 \pm 10 \, \Omega \, °C^{-1}$ in the temperature range of 25 °C–45 °C [2].

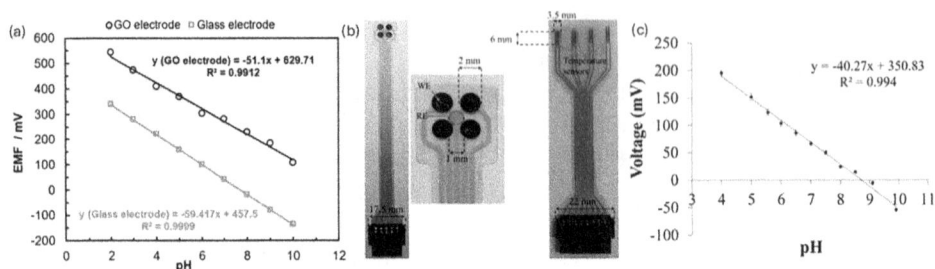

Figure 6.7. (a) Potentiometric pH sensing of GO coated on graphitic carbon electrode. Reproduced from [32] with permission from Springer Nature. (b) Image of a GO-based pH sensor integrated with a printed temperature sensor, reprinted from [2], copyright (2017), with permission from Elsevier. (c) Potentiometric pH sensing performances of a printed GO-based sensor, reprinted from [2], copyright (2017), with permission from Elsevier.

The oxygen-functional groups present in the reduced GO play a significant role in the pH sensing. For an ion-to-electron transducer in pH sensing, the modified GO will be highly beneficial due to the enhancement of electrical conductivity and mechanical strength. The rGO is normally prepared by chemical, thermal or electrochemical reduction approach. The preparation of various forms of reduced GO from graphite is shown in figure 6.8 [33], in which the oxidation of graphite with chemicals leads to generation of graphite oxide (graphite-OX). The thermal reduction/exfoliation of graphite-OX leads to generation of thermally reduced graphene oxide (TR-GO). The exfoliation of graphite-OX by ultrasonication generates graphene oxide (graphene-OX or GO). The chemical reduction of GO leads to form chemically reduced GO (CR-GO), and the electrochemical reduction leads to electrochemically reduced GO (ER-GO) [33]. Detailed electrochemical properties of various GOs are discussed in one of the reported works [33].

It was found that the presence of oxygen-functional groups is highest in hydro-thermal reduced GO (HRGO) as compared to CR-GO and ER-GO [33]. However, the pH sensing studies were carried out for the ER-GO-based composite structure, including ER-GO with IRO_2 and ER-GO with polymers [34, 35].

6.5 Carbon composites for pH sensing

In one of the works, carbon–polyaniline composite structure was prepared for a stretchable potentiometric pH sensor. Direct laser carbonisation was employed to fabricate a high-performance stretchable pH sensor. The PANI is spray-coated on the carbon structure to make a pH-sensitive electrode [36]. The fabrication process is shown in figure 6.9(a) [36]. The laser-induced generation of conductive carbon micro- and nanomaterials exhibits high conductivity and stretchability. Here, the reference electrode Ag/AgCl was prepared by printing of Ag paste on one side of the PANI/ carbon layer and Ag layer treated with $FeCl_3$ solution to convert Ag/AgCl. The printed electrodes show multiple stretchable patterns, as shown in figure 6.9(b) [36]. The sensor exhibits excellent potentiometric performance (as shown in figure 6.9(c))

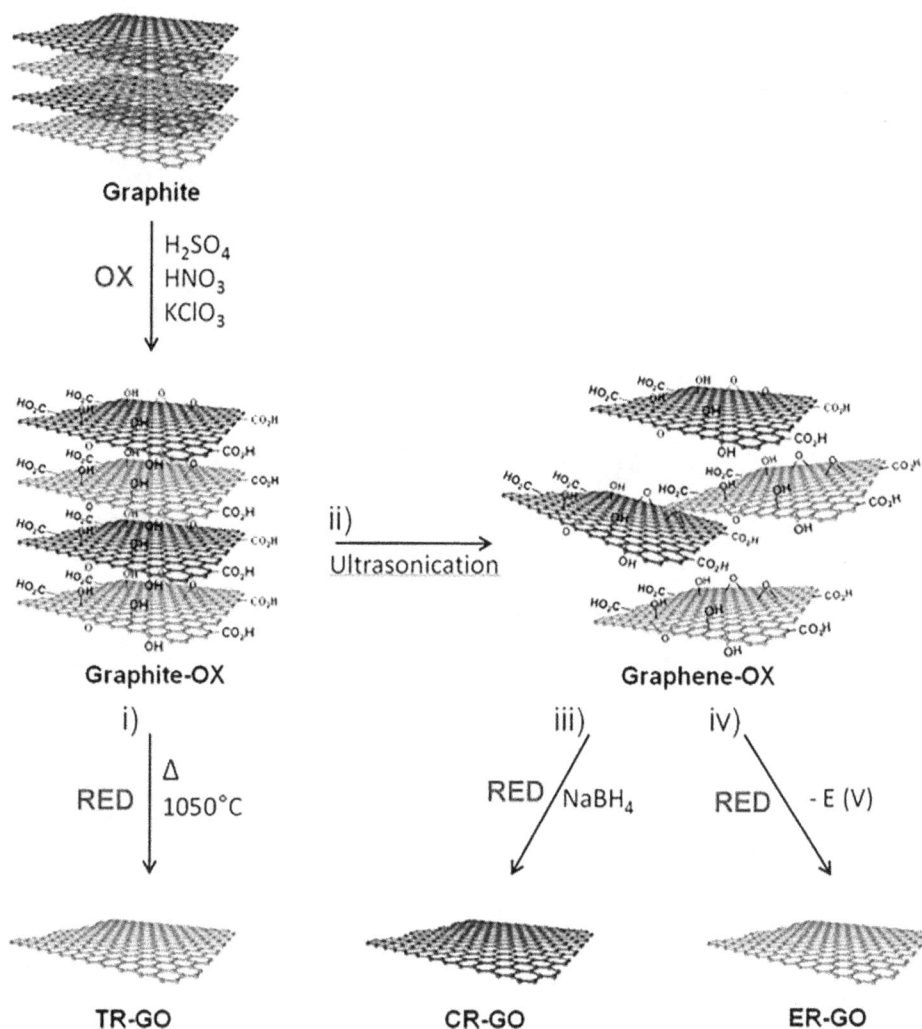

Figure 6.8. Schematic representation of chemically modified graphene for the preparation of various GOs [33]. John Wiley & Sons. Copyright 2011 WILEY-VCH Verlag GmbH & Co. KGaA, Weinheim.

with a sensitivity of 53 mV/pH in the range of pH 4–10. In addition to this, the developed sensor shows excellent longitudinal (up to 100%) and transverse strain (up to 100%). In both conditions, the variation of output value is less than ±5 mV. The image of longitudinal strain of the full sensor and its potential output is shown in figures 6.9(d) and (e) [36]. In another work, it was found that for the deposition of PANI, the carbon fibre offers excellent mechanical properties and a 2D conducting framework. For their fabrication, additional binders are not required and it leads to a simple method of development [37]. For a stretchable pH sensor and also textile-based pH sensor fabrication for sweat monitoring, it was found that graphite-polyurethane composites also exhibit excellent mechanical, electrical and electrochemical properties.

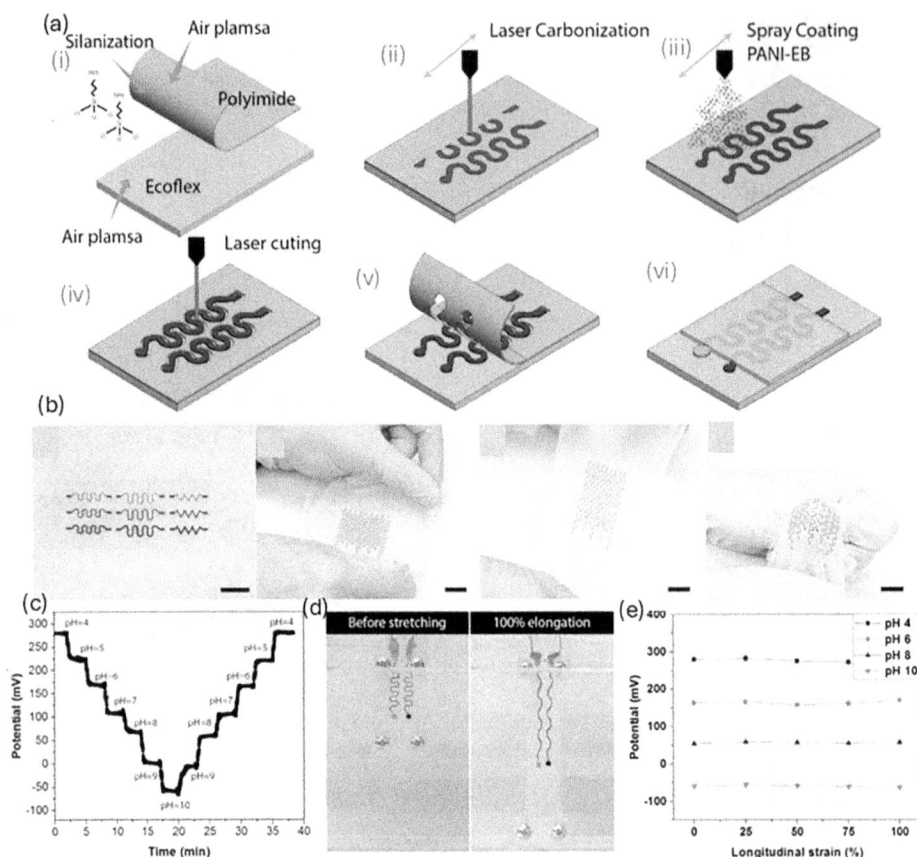

Figure 6.9. (a) Schematic illustrations of the fabrication process PANI-carbon stretchable pH sensor (i) polyimide sheet is silanized and placed on an air-plasma-treated ecoflex substrate, (ii) a CO_2 laser is used to carbonize serpentine carbon traces on the polyimide sheet, (iii) polyaniline is spray-coated onto the porous carbon, (iv) the polyimide sheet is machined with the same CO_2 laser at a higher power level, (v) excess polyimide is removed, (vi) interconnects are insulated by another Ecoflex layer followed by the deposition of Ag/AgCl and solid electrolyte, (b) photograph of various stretchable PANI/C–PI interconnect designs, (c) potentiometric performances of the pH sensor, (d) image of longitudinal strain of the sensor, and (e) potentiometric performances of the sensor under longitudinal strain. Reprinted with permission from [36], copyright (2017) American Chemical Society.

Details of these works are given in chapter 9. There are many works reported on the carbon-polymer composites, and they show advanced applications for wearables and environmental applications [36, 38–44].

6.6 Summary

In an electrochemical sensor, the carbon-based electrodes offer high electrical conductivity, high specific surface area, low weight and small sizes, high porosity, excellent chemical and thermal stability and low cost and mass production possibilities. These unique properties have lead to significant studies of researchers in carbon-based electrodes for pH sensors. This chapter provides an overview of various carbon nanostructures implemented for pH sensing. The sensing performances and mechanism of carbon-based pH sensing are provided in this chapter.

References

[1] Kangmennaa A, Forkuo R B and Agorku E S 2024 Carbon-based electrode materials for sensor application: a review *Sens. Technol.* **2** 2350174

[2] Salvo P *et al* 2017 Temperature and pH sensors based on graphenic materials *Biosens. Bioelectron.* **91** 870–7

[3] Goh G L, Agarwala S, Tan Y J and Yeong W Y 2018 A low cost and flexible carbon nanotube pH sensor fabricated using aerosol jet technology for live cell applications *Sens. Actuators* B **260** 227–35

[4] Kaur H, Siwal S S, Chauhan G, Saini A K, Kumari A and Thakur V K 2022 Recent advances in electrochemical-based sensors amplified with carbon-based nanomaterials (CNMs) for sensing pharmaceutical and food pollutants *Chemosphere* **304** 135182

[5] Fantinelli Franco F, Malik M H, Manjakkal L, Roshanghias A, Smith C J and Gauchotte-Lindsay C 2024 Optimizing carbon structures in laser-induced graphene electrodes using design of experiments for enhanced electrochemical sensing characteristics *ACS Appl. Mater. Interfaces* **16** 65489–502

[6] Joshi P, Mishra R and Narayan R J 2021 Biosensing applications of carbon-based materials *Curr. Opin. Biomed. Eng.* **18** 100274

[7] Channabasavana Hundi Puttaningaiah K P 2024 Innovative carbonaceous materials and metal/metal oxide nanoparticles for electrochemical biosensor applications *Nanomaterials* **14** 1890

[8] Wang L, Gu C, Wu L, Tan W, Shang Z, Tian Y and Ma J 2024 Recent advances in carbon dots for electrochemical sensing and biosensing: a systematic review *Microchem. J.* **207** 111687

[9] Power A C, Gorey B, Chandra S and Chapman J 2018 Carbon nanomaterials and their application to electrochemical sensors: a review *Nanotechnol. Rev.* **7** 19–41

[10] Gibi C, Liu C-H, Barton S C, Anandan S and Wu J J 2024 Carbon materials for electrochemical sensing application—a mini review *J. Taiwan Inst. Chem. Eng.* **154** 105071

[11] Villarreal C C, Pham T, Ramnani P and Mulchandani A 2017 Carbon allotropes as sensors for environmental monitoring *Curr. Opin. Electrochem.* **3** 106–13

[12] Kirchner E-M and Hirsch T 2020 Recent developments in carbon-based two-dimensional materials: synthesis and modification aspects for electrochemical sensors *Microchim. Acta* **187** 441

[13] Selvam A, Sharma R, Sutradhar S and Chakrabarti S 2021 Synthesis of carbon allotropes in nanoscale regime *Carbon Nanomaterial Electronics: Devices and Applications* ed A Hazra and R Goswami (Singapore: Springer) pp 9–46

[14] Ahmad F, Mahmood A and Muhmood T 2024 Carbon allotropes: basics, properties and applications *Heteroatom-Doped Carbon Allotropes: Progress in Synthesis, Characterization, and Applications* (Washington, DC: American Chemical Society) pp 1–18

[15] Krasley A T, Li E, Galeana J M, Bulumulla C, Beyene A G and Demirer G S 2024 Carbon nanomaterial fluorescent probes and their biological applications *Chem. Rev.* **124** 3085–185

[16] Dai H 2002 Carbon nanotubes: opportunities and challenges *Surf. Sci.* **500** 218–41

[17] Zhu L, Yang R, Zhai J and Tian C 2007 Bienzymatic glucose biosensor based on co-immobilization of peroxidase and glucose oxidase on a carbon nanotubes electrode *Biosens. Bioelectron.* **23** 528–35

[18] Santhosh P, Manesh K, Gopalan A and Lee K-P 2007 Novel amperometric carbon monoxide sensor based on multi-wall carbon nanotubes grafted with polydiphenylamine—fabrication and performance *Sens. Actuators* B **125** 92–9

[19] Ahammad A J S, Lee J J and Rahman M A 2009 Electrochemical sensors based on carbon nanotubes *Sensors (Basel)* **9** 2289–319

[20] Lei K F, Lee K-F and Yang S-I 2012 Fabrication of carbon nanotube-based pH sensor for paper-based microfluidics *Microelectron. Eng.* **100** 1–5

[21] Abdulhameed A, Halin I A, Mohtar M N and Hamidon M N 2021 pH-Sensing characteristics of multi-walled carbon nanotube assembled across transparent electrodes with dielectrophoresis *IEEE Sens. J.* **21** 26594–601

[22] Liu L, Shao J, Li X, Zhao Q, Nie B, Xu C and Ding H 2016 High performance flexible pH sensor based on carboxyl-functionalized and DEP aligned SWNTs *Appl. Surf. Sci.* **386** 405–11

[23] Pumera M, Ambrosi A, Bonanni A, Chng E L K and Poh H L 2010 Graphene for electrochemical sensing and biosensing *TrAC, Trends Anal. Chem.* **29** 954–65

[24] Salvo P, Melai B, Calisi N, Paoletti C, Bellagambi F, Kirchhain A, Trivella M G, Fuoco R and Di Francesco F 2018 Graphene-based devices for measuring pH *Sens. Actuators* B **256** 976–91

[25] Angizi S, Yu E Y C, Dalmieda J, Saha D, Selvaganapathy P R and Kruse P 2021 Defect engineering of graphene to modulate pH response of graphene devices *Langmuir* **37** 12163–78

[26] Ang P K, Chen W, Wee A T S and Loh K P 2008 Solution-gated epitaxial graphene as pH sensor *JACS* **130** 14392–3

[27] Ohno Y, Maehashi K, Yamashiro Y and Matsumoto K 2009 Electrolyte-gated graphene field-effect transistors for detecting pH and protein adsorption *Nano Lett.* **9** 3318–22

[28] Qi X, Jin W, Tang C, Xiao X, Li R, Ma Y and Ma L 2025 pH monitoring in high ionic concentration environments: performance study of graphene-based sensors *Anal. Sci.* **41** 127–35

[29] Lei N, Li P, Xue W and Xu J 2011 Simple graphene chemiresistors as pH sensors: fabrication and characterization *Meas. Sci. Technol.* **22** 107002

[30] Pei S and Cheng H-M 2012 The reduction of graphene oxide *Carbon* **50** 3210–28

[31] Chung C, Kim Y-K, Shin D, Ryoo S-R, Hong B H and Min D-H 2013 Biomedical applications of graphene and graphene oxide *ACC. Chem. Res.* **46** 2211–24

[32] Neupane S, Subedi V, Thapa K K, Yadav R J, Nakarmi K B, Gupta D K and Yadav A P 2022 An alternative pH sensor: graphene oxide-based electrochemical sensor *Emerg. Mater.* **5** 509–17

[33] Ambrosi A, Bonanni A, Sofer Z, Cross J S and Pumera M 2011 Electrochemistry at chemically modified graphenes *Chem.—Eur. J.* **17** 10763–70

[34] Yang J, Kwak T J, Zhang X, McClain R, Chang W-J and Gunasekaran S 2016 Digital pH test strips for in-field pH monitoring using iridium oxide-reduced graphene oxide hybrid thin films *ACS Sens.* **1** 1235–43

[35] Chinnathambi S and Euverink G J W 2018 Polyaniline functionalized electrochemically reduced graphene oxide chemiresistive sensor to monitor the pH in real time during microbial fermentations *Sens. Actuators* B **264** 38–44

[36] Rahimi R, Ochoa M, Tamayol A, Khalili S, Khademhosseini A and Ziaie B 2017 Highly stretchable potentiometric pH sensor fabricated via laser carbonization and machining of carbon—polyaniline composite *ACS Appl. Mater. Interfaces* **9** 9015–23

[37] Hossain M S, Padmanathan N, Badal M M R, Razeeb K M and Jamal M 2024 Highly sensitive potentiometric pH sensor based on polyaniline modified carbon fiber cloth for food and pharmaceutical applications *ACS Omega.* **9** 40122–33

[38] Mahinnezhad S, Emami H, Ketabi M, Shboul A A, Belkhamssa N, Shih A and Izquierdo R 2021 Fully printed pH sensor based in carbon black/polyaniline nanocomposite *2021 IEEE Sensors* pp 1–4

[39] Saikrithika S and Kumar A S 2021 A selective voltammetric pH sensor using graphitized mesoporous carbon/polyaniline hybrid system *J. Chem. Sci.* **133** 46

[40] Kaempgen M and Roth S 2006 Transparent and flexible carbon nanotube/polyaniline pH sensors *J. Electroanal. Chem.* **586** 72–6

[41] Dang W, Manjakkal L, Navaraj W T, Lorenzelli L, Vinciguerra V and Dahiya R 2018 Stretchable wireless system for sweat pH monitoring *Biosens. Bioelectron.* **107** 192–202

[42] Manjakkal L, Dang W, Yogeswaran N and Dahiya R 2019 Textile-based potentiometric electrochemical pH sensor for wearable applications *Biosensors* **9** 14

[43] Dang W, Manjakkal L, Lorenzelli L, Vinciguerra V and Dahiya R 2017 Stretchable pH sensing patch in a hybrid package *2017 IEEE Sensors* pp 1–3

[44] Hosseini E S, Manjakkal L and Dahiya R 2021 Flexible and printed potentiometric pH sensor for water quality monitoring *2021 IEEE Int. Conf. on Flexible and Printable Sensors and Systems (FLEPS)* pp 1–4

Chapter 7

Conducting polymers-based pH sensors: sensor fabrication and mechanism

7.1 Introduction

In electrochemical pH sensing, there are several materials and methods implemented for the development of a sensor. The biocompatibility, flexibility, and wearability, as well as the customizable electrochemical and electrical properties, and the possibility of miniaturised electrode design, are highly needed for pH sensing fabrications in various applications, including biomedical and pollution monitoring. Recently advanced pH sensors were found to be importances in wound monitoring, body fluid analysis and other implantable applications [1]. The conventional pH glass electrodes are not suitable for wearable systems due to a lack of bending capability, and glass could easily crack during user movement [2]. Further, the glass pH-sensitive electrodes are large, difficult to miniaturise, and require regular topping up of the reference buffer solution [3–5]. As a result, alternatives such as metal oxides and carbon-based electrodes are used for pH sensor development [6]. Besides the possibility of miniaturisation and flexibility, they offer several attractive features such as faster response, wider pH sensing range, excellent sensitivity, simple electronics, biocompatibility, low cost of fabrication, and the possibility of integration on different substrates (polymer, plastic, textiles, paper, etc). However, the lack of stability of carbon-based electrodes and the issues related to oxide dissolution of the electrodes in highly alkaline or highly acidic media, of metal oxides, jave lead to the search for an alternative pH-sensitive electrode [7, 8]. For this electrochemical pH sensing, the conducting polymers are identified as an excellent material due to their unique properties [9]. The conducting polymers have received excellent attention in various electrochemical sensing applications. A summary of various conducting polymers, their characteristics, the main preparation methods and potential applications are summarised in figure 7.1.

doi:10.1088/978-0-7503-6079-1ch7
7-1

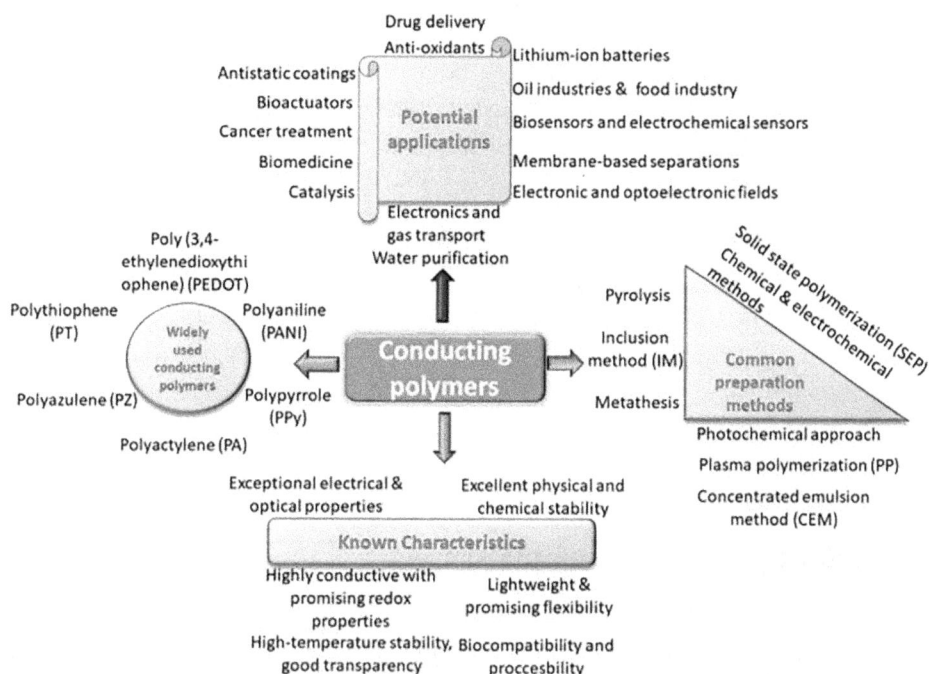

Figure 7.1. Overview of various conducting polymers, their preparation methods, characteristics and potential applications. Reprinted from [9], copyright (2025), with permission from Elsevier.

Conducting polymers exhibit excellent pH sensitivity due to the protonation/deprotonation of functional groups within the materials, which is dependent on the pH value of the solution [10, 11]. In addition, the polymers could be used as a general matrix and further modified for selective sensing [12]. Polyaniline (PANI), poly(3,4-ethylenedioxythiophene) (PEDOT) and polypyrrole (PPy) are the major studied conducting polymers implemented for pH sensing due to their metallic-like conductivity. The conductivity of different polymers is compared in figure 7.2 [13] along with other materials. The conducting polymers can be classified as n-doping or p-doping. In p-doping, the electrons move from the highest occupied molecular orbital (HOMO) of the polymer to the dopant, resulting in a loss of electrons [13]. In n-doping, electrons from the dopant migrate to the lowest unoccupied molecular orbital (LUMO) of the polymer. This leads to enhancing the electrode density and enabling electron conduction [13]. The conducting polymers are both ionic and electronic conductors and can easily act as transducers for various ionic or molecular signals. Based on their high conductivity and electrochemical properties, conducting polymers have excellent applications in energy storage, electrochemical sensing and biomedicine [13–21].

For the next generation of biocompatible pH sensing, the conductive polymers will be ideal materials. A detailed investigation of these materials' pH sensing will enable evaluation of their applicability. This chapter discusses the major progress and mechanism of conductive polymer-based electrochemical pH sensing.

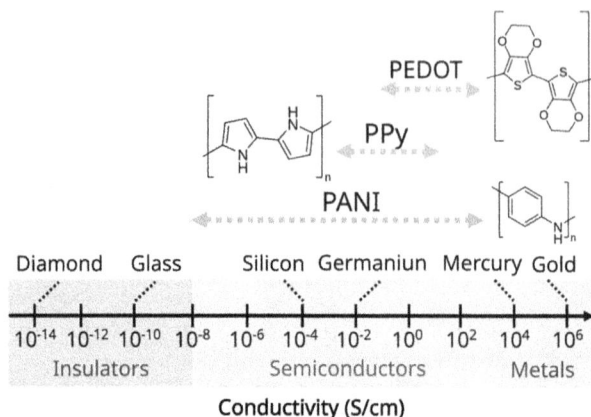

Figure 7.2. Comparison of the conductivity range of various polymers with other materials. Reproduced from [13] CC BY 4.0.

7.2 Various polymers-based pH sensors

7.2.1 Polyaniline-based pH sensor

Polyaniline (PANI) is one of the excellent conducting polymers that has been studied extensively [1]. PANI exhibits high conductivity, low energy optical transitions, low ionization potential and large surface area for its nanostructures [22]. The main advantages of PANI is its easy preparation method, low cost, high stability, processability and excellent mechanical properties. The doping and de-doping of the PANI molecular chain control the conductivity of the PANI film. The molecular structure of PANI is shown in figure 7.3(a), and it consists of a series of reduced and oxidised structural units [23]. Here, the y represents the degree of reduction of PANI. PANI exists in three different base forms: leucoemeraldine (LEB, fully reduced, where $y = 1$); emeraldine (EB, half-oxidised, where $y =$ between 0 and 1, in which when $y = 0.5$ the dopant is alternatively doped in the molecular chain and PANI is in an intermediate oxidation state); and pernigraniline (PNB, fully oxidised, when $y = 0$) [23, 24]. In the PANI-EB state, the conductivity of the PANI film is optimal. In general conditions, PANI is a mixture of all these three states, and for high-performing electronic and electrochemical devices a high portion of EB is expected [23]. Due to their unique molecular structure and properties, PANI film shows excellent applications in energy storage (batteries and supercapacitors), displays, corrosion inhibitors and electrochemical sensors [24]. The electrically conductive form of PANI is the emeraldine salt (ES) form, which is the protonated form of EB. This EB conversion can easily be achieved with strong acids due to the presence of basic sites such as amine and imine groups in the polymer, as shown in figure 7.3(b) [24]. This protonation process is reversible and is also valid for alkyl-substituted PANIs [24]. These protonation and deprotonation abilities make PANI an ideal pH-sensitive material [25].

There are various approaches implemented for the development of PANI film for the sensor fabrication and in which the major works are reported for the flexible and

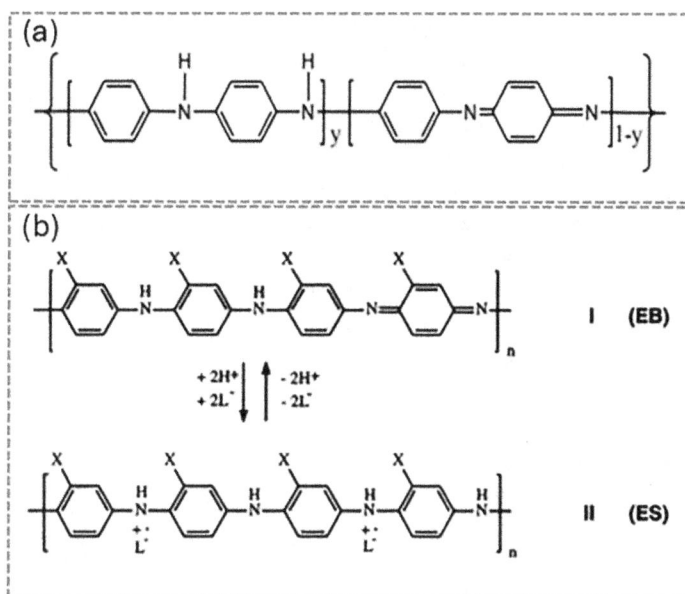

Figure 7.3. (a) Molecular structure of PANI, reproduced from [23] CC BY 4.0, and (b) schematic for PANI protonation and deprotonation, reprinted from [24], copyright (2002), with permission from Elsevier.

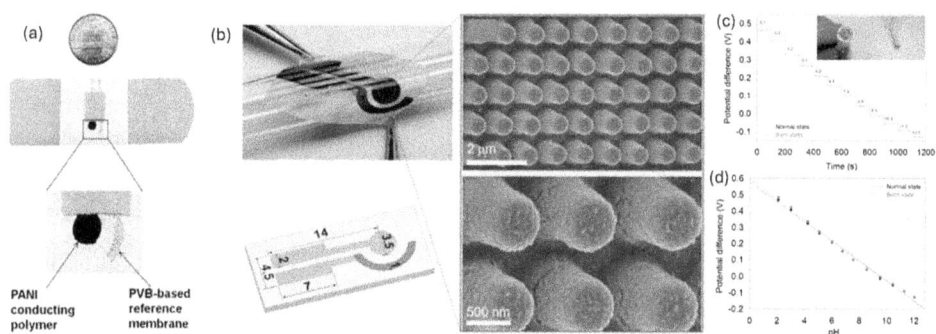

Figure 7.4. (a) Schematic representation of smart bandage-based pH sensor [26] John Wiley & Sons. Copyright 2014 WILEY-VCH Verlag GmbH & Co. KGaA, Weinheim. (b) Photograph and SEM of flexible PANI nanopillar array-based pH sensor. (c) and (d) Potentiometric pH sensing performance in the normal state and the bending state of the PANI nanopillar array-based pH sensor. (b)–(d) reprinted from [10], copyright (2017), with permission from Elsevier.

wearable applications. For a wearable application, the first bandage-based pH sensor was developed using PANI as pH-sensitive electrode, which was prepared via electropolymerization [26]. The electrodes were printed on the top of the bandage via screen printing. The schematic of the bandaged-based pH sensor is given in figure 7.4(a) [26] and the developed sensor was able to measure the pH in the range of 4.35–8 with a sensitivity of 58 mV/pH [26]. Detailed evaluation and performance of this flexible sensor are discussed in chapter 9.

The electrochemical surface area of the PANI film is achieved through the design of the PANI nanopillar array developed using a soft lithography approach, and it shows applications in disposable and flexible pH sensors. For this flexible sensor, the Ag/AgCl reference electrode was prepared on patterned-nanopillar backbone films by stencil lithography [10]. The image and schematic representation, along with a microscopic image, are shown in figure 7.4(b) [10]. The sensor provides a linear response of 60.3 mV/pH in a wide pH range 2–12. The sensor exhibits negligible performance in both with and without bending, as shown in figure 7.4(c) and (d) [10].

For the development of the PANI fibre network, the screen-printed electrodes offer an excellent platform. In one of the works, the PANI nanofibre arrays were coated on top of screen-printed carbon electrodes. The printing also allows for the fabrication of the reference electrode on the same substrate. To expand the active sites of the sensitive electrode for hydrogen ions interaction, a three-dimensional PANI (3D PANI) film was prepared [27]. The 3D porous PANI was constructed by electro-polymerisation on an interdigitated electrode (IDE) gold electrode. The schematic of the fabrication process is shown in figure 7.5(a) [27]. On this PANI electrode, a reversible transformation of proton gains and losses occurs between the emeraldine salt and the emeraldine base [27]. The sensors show excellent electro-chemical performance, with and without bending. The potentiometric performance

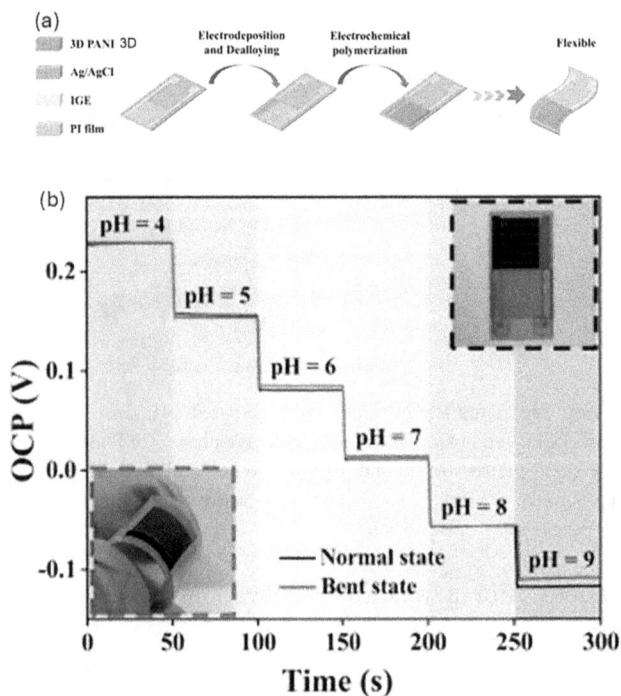

Figure 7.5. (a) 3D porous PANI-based pH sensor construction by electro-polymerisation on IDE, (b) The variation of open circuit potential with normal state and bending state of the pH sensor. Reprinted from [27], copyright (2023), with permission from Elsevier.

of the sensors is shown the figure 7.5(b) [27]. The 3D PANI sensor exhibited a super-Nernstian response with a sensitivity of 69.33 mV/pH in the range of pH 4–9 [27]. This electrode is implemented for the sweat pH sensing, and it shows a very fast response time of 7.75 s, reversibility and repeatability. The drift rate ($0.10 \, \text{mV h}^{-1}$) and influence of temperature ($0.06 \, \text{mV/pH/}^{\circ}\text{C}$) is significantly low, showing excellent applications in wearable devices [27]. The additive manufacturing, such as 3D printing, allows mass production, and to further reduce the cost it will be good to use carbon-based electrodes instead of gold IDE. Fine-tuning of the structural and morphological properties of this 3D electrode could potentially enhance the sensing performance. The 3D printing allows the formation of various porous structures in a 3D electrode for enhancing the electrochemically active surface area [16]. The 3D printed electrodes are well studied for portable and miniaturised electrochemical biosensors. To enhance the sensitivity of the PANI-based sensor, another approach utilised in sensor fabrication is the acid doping of PANI. The electropolymerized PANI film was processed with acid doping using concentrated sulfuric acid. Acid-doped PANI exhibits negligible hysteresis and promising stability [28]. It was found that the acid doping reveals a phosphate and PANI forms a salt, which makes PANI change from an insulator to a metal state and improves the conductivity, which leads to giving a super-Nernstian response ($-69.31 \pm 1.66 \, \text{mV/pH}$) for the sensor [28].

Another innovative design implemented for the development of stretchable pH sensor using PANI is the preparation of sensitive and reference electrodes on a gold fibre electrode. The PANI film and Ag layer are coated on the top of the gold fibre by the electrodeposition method, as shown in figure 7.6(a) [29]. Here, the Ag layer is obtained through the electrochemical deposition of Ag onto the gold fibre surface

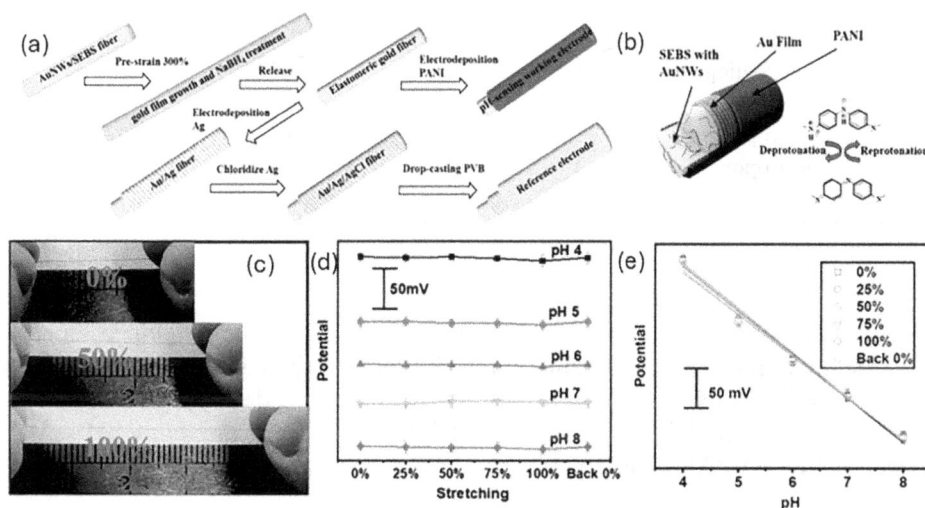

Figure 7.6. (a) Schematic of fibre-based pH sensing and reference electrodes fabrication, (b) the cross-section and mechanism of the PNAI electrode-based pH sensing, (c) image of stretchability of the sensor, (d) the potentiometric performances of the sensor in different pH solution under different stretching condition, and (e) the linear relationship between the sensor with different pH value under different stretching condition. Reproduced from [29] with permission from the Royal Society of Chemistry.

Table 7.1. Comparison of performances of various conducting polymers-based pH sensors.

Materials	Fabrication	pH range	Response time	Sensitivity (mV/pH)	References
PANi	Electrodeposition	1–13	A few seconds	58	[49]
PANi	Electrodeposition	4–7	~60 s	—	[50]
PANi	Drop-casting	4–10	Rise time 12 s Fall time 36 s (pH 6–8)	50–58.2 (pH 2–12)	[51]
PANi	Electropolymerization	5.5–8	20 s	58.0 ± 0.3	[52]
PANi	Electropolymerization	3–7	25 s	52.8–59.6 (dependent on bending/ stretching conditions)	[53]
PANi	Electrodeposition	2–12	<10 s	58.2	[54]
PANi	Dilute chemical polymerization	3.9–10.1	12.8 s	62.4	[55, 56]
PPy	Electropolymerization	2–12	<10 s	54.67 ± 0.7	[57]
PPy + CNT, PANI + CNT	Electrodeposition	1–13		59.2	[49]
PANI + PPy + PSS	Inkjet printing	3–10		60.6	[46]
PANI + DBSA	Spin coating	5.4–8.6		58.57	[58]
PPy	Electropolymerization	3–10		46	[40]
PAA-CNTs	Electropolymerization	2–12	3 s	54.5	[59]

through cyclic voltammetry, as shown in figure 7.6(b) [29]. The developed sensor is able to stretch up to 100% with minimal deviation (less than 3%) in electrochemical performance, as shown in figures 7.6(c) and (d) [29]. The developed sensor shows sensitivity of 60.6 mV/pH in the range of pH 4–8 [29]. There are many pH sensors reported based on PANI electrode, and a few of them are reported in table 7.1.

7.2.2 Polypyrrole-based pH sensor

PPy is one of the most extensively studied conductive polymers for electrochemical and biosensing [30]. It exhibits a high level of conductivity, redox activity, environmental stability, and tuneable physicochemical properties and morphologies [30]. Due to these properties, PPy shows excellent applications in electrochemical devices, including sensors and energy storage [30–33]. The chemical and electrochemical sensing performance of PPy is based on a reversible change of conductivity from insulator to metal caused by a doping/de-doping process [34]. The various structures of PPy were employed for the development of various types of electrochemical pH sensors, including thin-film microelectrodes, IDE-based sensors, nanowires, and nanopillar array-based potentiometric sensors [30, 34–36].

For low-cost and disposable applications, the PPy-based pH sensors developed on paper substrates show precision of measurement in 0.02–0.13 pH units in a pH range of 8–10 [37]. For mass production of the disposable pH sensor, one of the approaches used is the design of sensors on screen-printed carbon electrodes (SPCEs) in a textile substrate [38]. The PPy was used as a sensitive electrode on this SPCE, and it shows a slope value of 54.173 mV/decade [38]. Due to the single use of sensors, one parameter that needs to be considered in developing polymer-based pH sensors is the low manufacturing technology. For flexible electronics fabrication, one of the low-cost methods for mass production is inkjet printing. Inkjet printing allows fabrication of disposable, robust and miniaturised sensors. One of the excellent works reported on printed conducting polymer-based pH sensors is the design of PANI: PSS/PPy: PSS electrode by inkjet printing [39]. The fabrication steps of the sensor are shown in figure 7.7(a) [39]. The fabrication

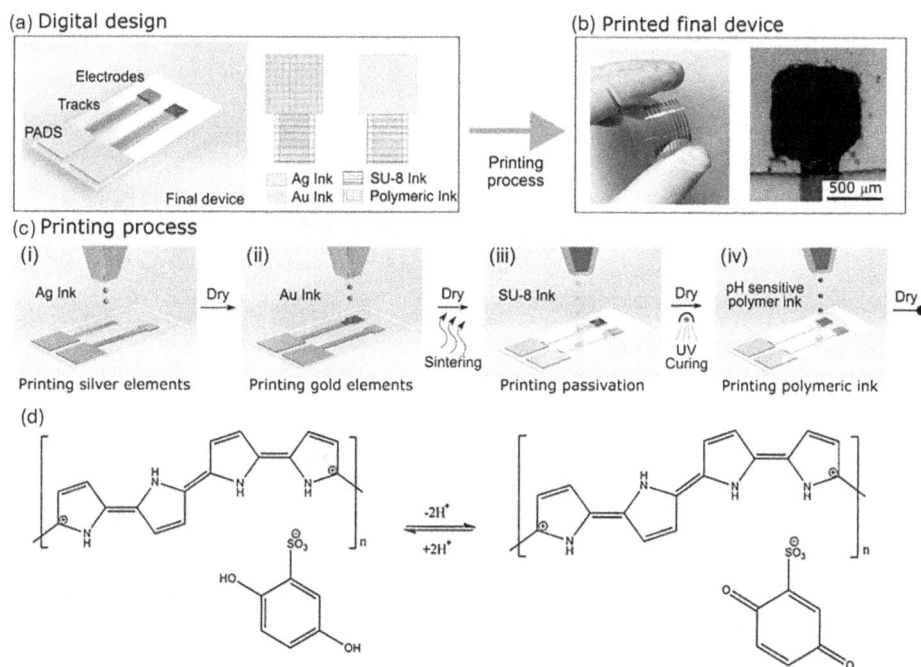

Figure 7.7. (a) Digital design, (b) printed sensors and (c) fabrication steps of inkjet printed pH sensor, reproduced from [39] CC BY 4.0, and (d) protonation and deprotonation of HQS-PPy, reprinted from [34], copyright (2013), with permission from Elsevier.

consists initially of a digital design of the electrodes and sensitive layers, and the fabricated sensors are given in figure 7.7(b) [39]. Figure 7.7(c) provides the detailed steps of the fabrication of the electrodes through printing. Here, Su-8 ink is the dielectric layer. The gold electrode layer supports the enhancement of electron mobility between the sensitive polymer layer and Au nanoparticles [39]. This sensor shows a sensitivity of 81.05 ± 0.08 mV/pH and it reveals a super-Nernstian response [39].

To enhance the sensing performance a PPy nanowire array was developed through anodic aluminium oxide (AAO) template-assisted process [34]. In this the PPy nanowire array was electropolymerized using hydroquinone monosulfonate (HQS) as a functional dopant. The HQS possesses a quinhydrone-like structure and is well studied for pH sensing due to its well-known molecular charge-transfer complex [34, 40]. The pH sensing mechanism such as protonation/ deprotonation of HQS-PPy is shown in figure 7.7(d) [34]. While comparing the potentiometric pH sensing performances, it was found that pH sensors based on PPy-HQS nanowires exhibited higher sensitivity (49.33 mV/pH) than those based on PPy-HQS thin films (3.01 mV/pH) [34].

The nanopillar or nanowire array structures effectively increase the specific surface area and accelerate the interaction process with analytes for electrochemical sensing [41, 42]. One of the novel methods utilised for the nanopillar array of PPy for pH sensing is the aid of a porous track-etched membrane using O_2 plasma treatment as the etching technique, as shown in figure 7.8(a) [42]. PPy nanopillar array-based electrodes exhibit a more hydrophilic nature and hence the protonation/deprotonation process is more active. Due to these properties, the PPy nanopillar array-based sensor shows excellent pH sensitivity (61 mV/pH), as shown in figure 7.8(b), and repeatability [42]. The developed antimicrobial PPy nanopillar array is ideal for

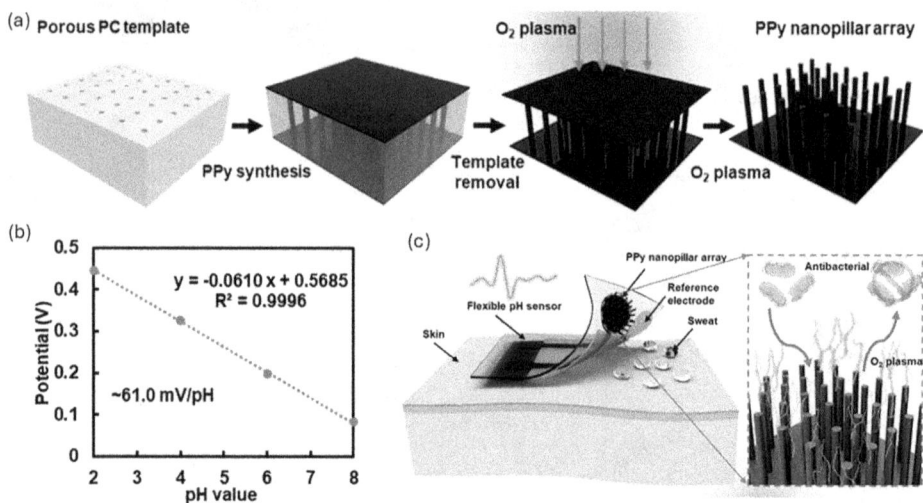

Figure 7.8. (a) Fabrication steps for antimicrobial PPy nanopillar array. (b) Potentiometric performances of PPy nanopillar array (c). Schematic representation of application of antimicrobial nanopillar array for wearable sweat monitoring [42], John Wiley & Sons. Copyright 2024 Wiley-VCH GmbH.

flexible sweat monitoring, as schematically represented in figure 7.8(c) [42]. There are various works reported on PANI passed pH sensor development and few of them are discussed in the table 7.1.

7.2.3 PEDOT-based pH sensor

PEDOT exhibits a relatively high stability and varied conductivity (10^{-3}–10^3 S cm^{-1}) compared to PPy and PANI [43]. Combining PEDOT with poly(4-styrene-sulfonate) (PSS) makes PEDOT: PSS, which is an attractive material for organic electronics and sensing [44]. The chemical structure of PEDOT: PSS is shown in figure 7.9(a) [45]. The PEDOT: PSS is easily soluble in water, and its conductivity could be easily varied by using solvents (ethanol, glycerol, DMSO etc). Due to the excellent conductivity and electrochemical properties, PEDOT: PSS film acts as a channel material in an organic electrochemical transistor (OECT)-based sensor [43]. A schematic representation of OECT based on PEDOT: PSS channel is shown in figure 7.9(b) [46]. Based on this PEDOT, a flexible OECT was developed for wearable pH sensing. For pH signal transduction PEDOT: dye composites were synthesized using Bromothymol Blue (BTB) and Methyl Orange (MO) pH dyes as PEDOT counterions [46, 47]. The electrochemical pH sensing mechanism of PEDOT with BTB is shown in figure 7.9(c) [46]. Here, during the electrochemical

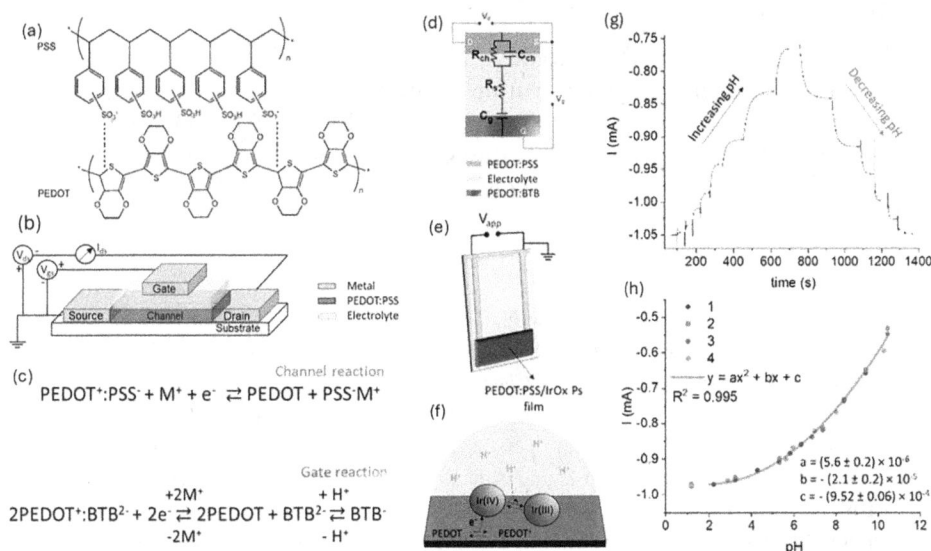

Figure 7.9. (a) Chemical structure of PEDOT:PSS, reproduced from [45] with permission from Springer Nature, (b) schematic of OECT-based pH sensor using PEDOT:PSS as a channel layer, (c) sensing mechanism of PEDOT with BTB based pH sensor, (d) ionic circuit of the OECT pH sensor. (b)–(d) reprinted with permission from [46], copyright (2018) American Chemical Society. (e) Two-electrode configuration based PEDOT:PSS/IrOx film, (f) sensing mechanism of the two-electrode configuration based pH sensor, (g) the variation of generated current with increasing and decreasing of pH values in a two-electrode sensor, and (h) polynomial curve describing the sensor response obtained from the four independent sensor data. (e)–(h) reproduced from [48] CC BY 4.0.

reaction, the protonation/deprotonation of the counterion BTB impacts on the potential and thus on the capacitance of PEDOT at the gate, eventually imparting the observed pH-dependency to the modulation of the drain current (I_d) [46]. The ionic circuit of the OECT pH sensor is shown in figure 7.9(d) [46]. This developed OECT pH sensor exhibits a pH sensitivity of 93 ± 8 mV/pH [46]. PEDOT: PSS is well studied for OECT-based pH sensing using PEDOT: PSS alone or with other composite/doped structures [47].

Another electrode configuration of PEDOT: PSS pH sensor implemented for wearable devices is the design of two-electrode configurations, as shown in figure 7.9(e) [48], in which the pH-sensitive electrode is developed based on chemically synthesized IrO_x particles (IrOx Ps) embedded in a PEDOT: PSS thin film. Here IrO_x enable it to reversibly change with pH variations due to spontaneous redox reactions. This electrode was developed on textile substrate and implemented for wound monitoring. The schematic representation of the sensor is shown in figure 7.9(e) and its sensing mechanism in figure 7.9(f) [48]. The increasing and decreasing of the generated current of the two-electrode sensor with changing pH value is shown in figure 7.9(g) [48]. The nonlinear response in pH range 2–11 is shown in figure 7.9(h) [48]. The studies reveal that the developed sensor exhibits a sensitivity of (59 ± 4) μA/pH and is medically relevant to the pH range for wound monitoring (pH 6–9) [48].

7.3 Summary

Conducting polymers exhibit excellent pH sensitivity due to the protonation/deprotonation of functional groups within the materials. Conducting polymer-based pH sensors have found significant interest in the design of flexible and wearable sensors in different electrode configurations. This chapter provides an overview of various conducting polymer-based pH sensors.

References

[1] Ghoneim M T, Nguyen A, Dereje N, Huang J, Moore G C, Murzynowski P J and Dagdeviren C 2019 Recent progress in electrochemical pH-sensing materials and configurations for biomedical applications *Chem. Rev.* **119** 5248–97

[2] Manjakkal L, Szwagierczak D and Dahiya R 2019 Metal oxides based electrochemical pH sensors: current progress and future perspectives *Prog. Mater. Sci.* **109** 100635

[3] Korostynska O, Arshak K, Gill E and Arshak A 2007 Materials and techniques for *in vivo* pH monitoring *IEEE Sens. J.* **8** 20–8

[4] Yuqing M, Jianrong C and Keming F 2005 New technology for the detection of pH *J. Biochem. Bioph. Methods* **63** 1–9

[5] Mostafalu P, Nezhad A S, Nikkhah M and Akbari M 2017 Flexible electronic devices for biomedical applications *Advanced Mechatronics and MEMS Devices II* (Berlin: Springer) pp 341–66

[6] Dang W, Manjakkal L, Navaraj W T, Lorenzelli L, Vinciguerra V and Dahiya R 2018 Stretchable wireless system for sweat pH monitoring *Biosens. Bioelectron.* **107** 192–202

[7] Lindino C A and Bulhões L O S 1996 The potentiometric response of chemically modified electrodes *Anal. Chim. Acta* **334** 317–22

[8] Alam A U, Qin Y, Nambiar S, Yeow J T W, Howlader M M R, Hu N-X and Deen M J 2018 Polymers and organic materials-based pH sensors for healthcare applications *Prog. Mater. Sci.* **96** 174–216

[9] Dube A, Malode S J, Alodhayb A N, Mondal K and Shetti N P 2025 Conducting polymer-based electrochemical sensors: progress, challenges, and future perspectives *Talanta Open* **11** 100395

[10] Yoon J H, Hong S B, Yun S-O, Lee S J, Lee T J, Lee K G and Choi B G 2017 High performance flexible pH sensor based on polyaniline nanopillar array electrode *J. Colloid Interface Sci.* **490** 53–8

[11] Talaie A 1997 Conducting polymer based pH detector: a new outlook to pH sensing technology *Polymer* **38** 1145–50

[12] Rahman M A, Kumar P, Park D-S and Shim Y-B 2008 Electrochemical sensors based on organic conjugated polymers *Sensors* **8** 118–41

[13] Gamboa J, Paulo-Mirasol S, Estrany F and Torras J 2023 Recent progress in biomedical sensors based on conducting polymer hydrogels *ACS Appl. Bio Mater.* **6** 1720–41

[14] Gualandi I, Tessarolo M, Mariani F, Possanzini L, Scavetta E and Fraboni B 2021 Textile chemical sensors based on conductive polymers for the analysis of sweat *Polymers* **13** 894

[15] Nair R, Uppuluri K, Paul F, Sirengo K, Szwagierczak D, Pillai S C and Manjakkal L 2025 Electrochemical energy storing performances of printed LaFeO$_3$ coated with PEDOT: PSS for hybrid supercapacitors *Chem. Eng. J.* **504** 158781

[16] Chandran A C S, Schneider J, Nair R, Bill B, Gadegaard N, Hogg R, Kumar S and Manjakkal L 2024 Enhancing supercapacitor electrochemical performance with 3D printed cellular PEEK/MWCNT electrodes coated with PEDOT: PSS *ACS Omega* **9** 33998–4007

[17] Peringath A R, Bayan M A H, Beg M, Jain A, Pierini F, Gadegaard N, Hogg R and Manjakkal L 2023 Chemical synthesis of polyaniline and polythiophene electrodes with excellent performance in supercapacitors *J. Energy Storage* **73** 108811

[18] Manjakkal L, Franco F F, Pullanchiyodan A, González-Jiménez M and Dahiya R 2021 Natural jute fibre-based supercapacitors and sensors for eco-friendly energy autonomous systems *Adv. Sustain. Syst.* **5** 2000286

[19] Manjakkal L, Pullanchiyodan A, Yogeswaran N, Hosseini E S and Dahiya R 2020 A wearable supercapacitor based on conductive PEDOT:PSS-coated cloth and a sweat electrolyte *Adv. Mater.* **32** 1907254

[20] Soni M, Bhattacharjee M, Manjakkal L and Dahiya R 2019 Printed temperature sensor based on graphene oxide/PEDOT:PSS *2019 IEEE Int. Conf. on Flexible and Printable Sensors and Systems (FLEPS)* pp 1–3

[21] Manjakkal L, Soni M, Yogeswaran N and Dahiya R 2019 Cloth based biocompatiable temperature sensor *2019 IEEE Int. Conf. on Flexible and Printable Sensors and Systems (FLEPS)* pp 1–3

[22] Shylendra S P, Wajrak M and Kang J J 2025 Advancements in solid-state metal pH sensors: a comprehensive review of metal oxides and nitrides for enhanced chemical sensing: a review *IEEE Sens. J.* **25** 7886–95

[23] Li Z and Gong L 2020 Research progress on applications of polyaniline (PANI) for electrochemical energy storage and conversion *Materials* **13** 548

[24] Lindfors T and Ivaska A 2002 pH sensitivity of polyaniline and its substituted derivatives *J. Electroanal. Chem.* **531** 43–52

[25] Yin Y, Guo C, Mu Q, Li W, Yang H and He Y 2024 Dual-sensing nano-yarns for real-time pH and temperature monitoring in smart textiles *Chem. Eng. J.* **500** 157115

[26] Guinovart T, Valdés-Ramírez G, Windmiller J R, Andrade F J and Wang J 2014 Bandage-based wearable potentiometric sensor for monitoring wound pH *Electroanalysis* **26** 1345–53

[27] Zhao Y, Yu Y, Zhao S, Zhu R, Zhao J and Cui G 2023 Highly sensitive pH sensor based on flexible polyaniline matrix for synchronal sweat monitoring *Microchem. J.* **185** 108092

[28] Bai Y, Zhu R, Zhao J and Cui G 2024 Super-Nernstian model based on acid doped polyaniline pH sensor *Microchem. J.* **203** 110715

[29] Wang R, Zhai Q, Zhao Y, An T, Gong S, Guo Z, Shi Q, Yong Z and Cheng W 2020 Stretchable gold fiber-based wearable electrochemical sensor toward pH monitoring *J. Mater. Chem.* B **8** 3655–60

[30] Ramanavičius A, Ramanavičienė A and Malinauskas A 2006 Electrochemical sensors based on conducting polymer—polypyrrole *Electrochim. Acta* **51** 6025–37

[31] Yuan L, Yao B, Hu B, Huo K, Chen W and Zhou J 2013 Polypyrrole-coated paper for flexible solid-state energy storage *Energy Environ. Sci.* **6** 470–6

[32] Dubal D P, Caban-Huertas Z, Holze R and Gomez-Romero P 2016 Growth of polypyrrole nanostructures through reactive templates for energy storage applications *Electrochim. Acta* **191** 346–54

[33] Jiao X and Liu Y 2025 Multifunctional polypyrrole-based flexible composite materials for next-generation smart material: integrated piezoresistive sensing, energy storage, electro-thermal heating, and UV protection *Composites* B **305** 112726

[34] Sulka G D, Hnida K and Brzózka A 2013 pH sensors based on polypyrrole nanowire arrays *Electrochim. Acta* **104** 536–41

[35] Lakard B, Segut O, Lakard S, Herlem G and Gharbi T 2007 Potentiometric miniaturized pH sensors based on polypyrrole films *Sens. Actuators* B **122** 101–8

[36] Carquigny S, Segut O, Lakard B, Lallemand F and Fievet P 2008 Effect of electrolyte solvent on the morphology of polypyrrole films: application to the use of polypyrrole in pH sensors *Synth. Met.* **158** 453–61

[37] Yue F, Ngin T S and Hailin G 1996 A novel paper pH sensor based on polypyrrole *Sens. Actuators* B **32** 33–9

[38] Sudarma A F, Alva S, Pangestu D and Hendri 2024 Development of disposal SPCE pH sensor based on polypyrrole and cloth as conductive polymer *Mater. Today Proc.* https://doi.org/10.1016/j.matpr.2024.03.018

[39] Zea M, Texidó R, Villa R, Borrós S and Gabriel G 2021 Specially designed polyaniline/polypyrrole ink for a fully printed highly sensitive pH microsensor *ACS Appl. Mater. Interfaces* **13** 33524–35

[40] Aquino-Binag C N, Kumar N, Lamb R N and Pigram P J 1996 Fabrication and characterization of a hydroquinone-functionalized polypyrrole thin-film pH sensor *Chem. Mater.* **8** 2579–85

[41] Xing J *et al* 2019 Antimicrobial peptide functionalized conductive nanowire array electrode as a promising candidate for bacterial environment application *Adv. Funct. Mater.* **29** 1806353

[42] Shang X, Chen P, Ling W and Hang T 2024 Controllable fabrication of polypyrrole nanopillar/hair arrays by oxygen plasma treatment and their applications as antibacterial flexible pH sensors *Adv. Mater. Technol.* **9** 2301992

[43] Gao N, Yu J, Tian Q, Shi J, Zhang M, Chen S and Zang L 2021 Application of PEDOT:PSS and its composites in electrochemical and electronic chemosensors *Chemosensors* **9** 79

[44] Reid D O, Smith R E, Garcia-Torres J, Watts J F and Crean C 2019 Solvent treatment of wet-spun PEDOT: PSS fibers for fiber-based wearable pH sensing *Sensors* **19** 4213

[45] Rahimzadeh Z, Naghib S M, Zare Y and Rhee K Y 2020 An overview on the synthesis and recent applications of conducting poly(3,4-ethylenedioxythiophene) (PEDOT) in industry and biomedicine *J. Mater. Sci.* **55** 7575–611

[46] Mariani F, Gualandi I, Tessarolo M, Fraboni B and Scavetta E 2018 PEDOT: dye-based, flexible organic electrochemical transistor for highly sensitive pH monitoring *ACS Appl. Mater. Interfaces* **10** 22474–84

[47] Mariani F, Gualandi I, Tonelli D, Decataldo F, Possanzini L, Fraboni B and Scavetta E 2020 Design of an electrochemically gated organic semiconductor for pH sensing *Electrochem. Commun.* **116** 106763

[48] Mariani F, Serafini M, Gualandi I, Arcangeli D, Decataldo F, Possanzini L, Tessarolo M, Tonelli D, Fraboni B and Scavetta E 2021 Advanced wound dressing for real-time pH monitoring *ACS Sens.* **6** 2366–77

[49] Kaempgen M and Roth S 2006 Transparent and flexible carbon nanotube/polyaniline pH sensors *J. Electroanal. Chem.* **586** 72–6

[50] Lee H, Song C, Hong Y S, Kim M S, Cho H R, Kang T, Shin K, Choi S H, Hyeon T and Kim D-H 2017 Wearable/disposable sweat-based glucose monitoring device with multistage transdermal drug delivery module *Sci. Adv.* **3** e1601314

[51] Rahimi R, Ochoa M, Parupudi T, Zhao X, Yazdi I K, Dokmeci M R, Tamayol A, Khademhosseini A and Ziaie B 2016 A low-cost flexible pH sensor array for wound assessment *Sensors Actuators* B **229** 609–17

[52] Guinovart T, Valdés-Ramírez G, Windmiller J R, Andrade F J and Wang J 2014 Bandage-based wearable potentiometric sensor for monitoring wound pH *Electroanalysis* **26** 1345–53

[53] Bandodkar A J, Hung V W S, Jia W, Valdés-Ramírez G, Windmiller J R, Martinez A G, Ramírez J, Chan G, Kerman K and Wang J 2013 Tattoo-based potentiometric ion-selective sensors for epidermal pH monitoring *Analyst* **138** 123–8

[54] Yoon J H, Kim K H, Bae N H, Sim G S, Oh Y-J, Lee S J, Lee T J, Lee K G and Choi B G 2017 Fabrication of newspaper-based potentiometric platforms for flexible and disposable ion sensors *J. Colloid Interface Sci.* **508** 167–73

[55] Park H J, Yoon J H, Lee K G and Choi B G 2019 Potentiometric performance of flexible pH sensor based on polyaniline nanofiber arrays *Nano Converg.* **6** 9

[56] Dahiya R, Yogeswaran N, Liu F, Manjakkal L, Burdet E, Hayward V and Jörntell H 2019 Large-area soft e-skin: the challenges beyond sensor designs *Proc. IEEE* **107** 2016–33

[57] Prissanaroon-Ouajai W, Pigram P J, Jones R and Sirivat A 2009 A sensitive and highly stable polypyrrole-based pH sensor with hydroquinone monosulfonate and oxalate co-doping *Sens. Actuators* B **138** 504–11

[58] Li Y, Mao Y, Xiao C, Xu X and Li X 2019 Flexible pH sensor based on a conductive PANI membrane for pH monitoring *RSC Adv.* **10** 21–8

[59] Gou P *et al* 2014 Carbon nanotube chemiresistor for wireless pH sensing *Sci. Rep.* **4** 4468

Chapter 8

Flexible and stretchable pH sensor: for e-skin in wearable devices and portable applications

8.1 Introduction

Electrochemical detection technologies for healthcare, environmental monitoring, food quality and other related fields have seen significant advancement in recent years due to their importance in our daily life [1–3]. Compared to traditional laboratory-based methods, the research is mainly focused on the design of sensors which can enable generation of data through online technology or an internet of things (IoT) system. For such real-time applications, the sensors need to be more mechanically flexible, especially when they need to be attached to curved surfaces, for example, in wearables. In wearables, along with flexibility, stretchability is also an important factor to consider when designing the sensors and related electronic components. The flexible sensors consist of a flexible substrate and active electrode materials which bend to any curved complex surfaces thus expanding the electro-chemical devices' applications to new enhanced domains. There are many nano-materials, based on inorganic, organic and their composite structures that are utilised for such sensor fabrication and a few of them are shown in figure 8.1(a) [1]. The lightweight structure, the possibility of implementation in portable applications and the biocompatibility of the materials enable the flexible electrochemical sensors to be applied in various applications, as summarised in figure 8.1(b) [1]. Based on these, there are different types of flexible electrochemical sensors reported for monitoring various parameters of body fluids, water, soil and health of animals [4–8].

The development of flexible and wearable pH sensors not only depends on the flexibility and sensing performance, but it also depends on the shape and attractive designs for various applications. In wearable devices, the sensors need to be either attached directly to skin or as clothing items, such as smart bandages or in textiles. Hence, along with flexibility, the wearability of the sensors is also highly important [5, 9, 10]. For continuous monitoring of pH value of body fluids, skin, brain tissue or

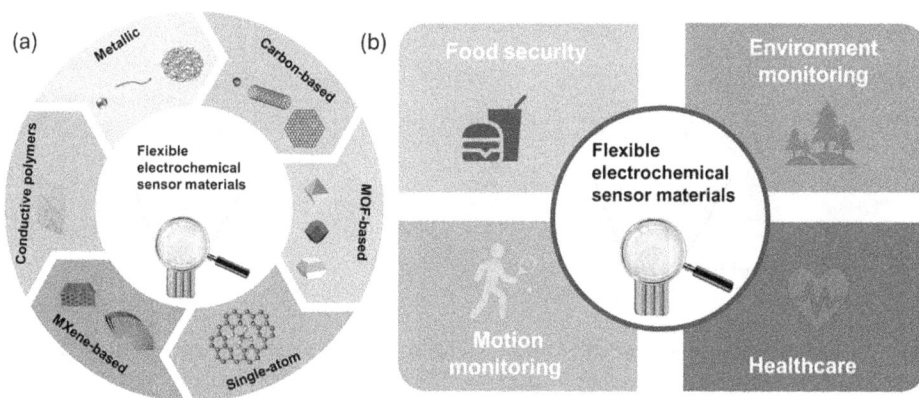

Figure 8.1. (a) Different types of materials used for flexible electrochemical sensors fabrication and (b) various applications of flexible electrochemical sensors. Reprinted from [1], copyright (2025), with permission from Elsevier.

other body parts, the sensors need to be integrated with suitable low-powered electronics and communication systems [11, 12]. Even though two-electrode-based potentiometric pH sensor exhibits high performance, wearable pH sensors are also developed using chemiresistive, field-effect transistors (FET) and three-electrode amperometric sensors. In addition to this, for water quality monitoring or other soil monitoring applications, the flexible pH sensors may need to be attached to curved surfaces, including unmanned remote vehicles for online monitoring purposes [2, 4, 13, 14].

This chapter discusses the most recent developments in flexible electrochemical potentiometric pH sensors, covering the key topics such as: (i) suitability of pH sensors in a wearable system; (ii) design of flexible pH sensors, which may vary with target applications; (iii) materials for various components of the sensor; and (iv) applications of flexible potentiometric pH sensors.

8.2 Flexible pH sensor designs

One of the key factors for the development of flexible and stretchable pH sensor for wearables is the choice of substrates. The substrate, which is often a flat, solid-state platform that facilitates the processing of sensitive materials, significantly influences the sensors' physical, mechanical, and electrical features [15, 16]. The choice of substrate also depends on the method of fabrication, processing temperature of the materials and biocompatibility. A detailed study of various substrates used for the flexible pH sensor design is provided in the previously reported work [5]. Some of the key properties of the substrate used for the deposition of sensitive and reference electrodes are given in figure 8.2.

In a potentiometric and amperometric sensor, design of the flexible reference electrode is also critical. Despite the challenging miniaturisation of thick- and thin-film reference electrodes, the development of a stable potential performing reference electrode is required for improving the overall performance of the pH sensor. The design of a flexible reference electrode using printing technology, especially

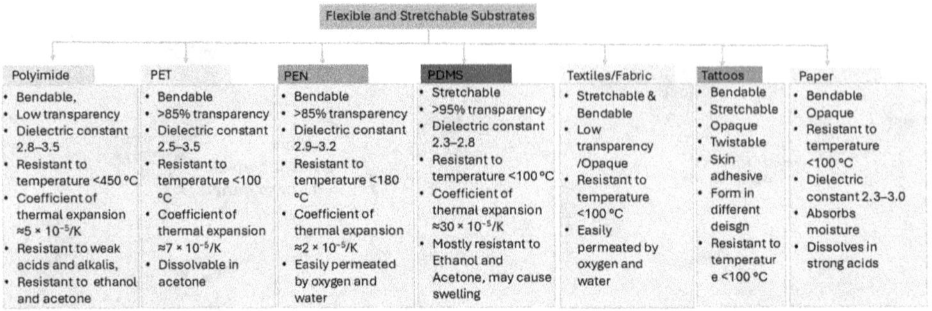

Figure 8.2. Substrates and their properties used for flexible and stretchable pH sensors fabrication (data from polyamide [17–21], polyethylene terephthalate (PET) [22–24], polyethylene naphthalene (PEN) [21, 25], polydimethylsiloxane (PDMS) [26–29], fibers, textiles and fabrics [30–32], tattoos [33, 34], paper [35, 36].

Ag|AgCl, is provided in chapter 3, and a few of the reported works are summarised in the review article [5].

8.3 Flexible pH-sensitive materials and sensors fabrication

Until now, different metal oxides, polymers and carbon nanomaterials have been used for the construction of electrochemical flexible pH sensors. Among these materials, in previous chapters we found that the metal oxides RuO_2 and IrO_2 were reported as an outstanding material for pH measurement over wide pH ranges, with fast, high sensitivity (close to Nernstian response), responses (less than 2 s), high accuracy and high durability. However, for flexible pH sensor applications, the implementation of these metal oxides is limited due to the requirement of low temperature processing and their flexibility or stretchability. In addition to this, for wearable devices, the non-toxicity and biocompatibility of the materials are highly important. For a wearable device recently, researchers have provided significant attention to preparing various ion-sensitive electrode materials, including composite structures and their different coating methods. Many of the prepared materials and sensors also found technical applications in monitoring different body parameters, including glucose, urea, Ca^{2+}, dopamine levels, etc, which are important for disease monitoring and diagnosis [37–44]. Researchers are now investigating a reliable pH measurement system for sweat, saliva, urine, wound fluid, tears and their real-time data collection. A detailed overview of various types of materials for wearable pH sensing is reported in multiple review articles [45] and a few of the key materials are discussed below.

In metal oxides, the IrO_2-based pH-sensitive electrode has garnered significant attention in the fabrication of flexible biomedical devices due to its excellent sensitivity and biocompatibility [46–48]. The flexible pH sensor fabricated by using IrO_2 has found application in brain tissue monitoring, measurement of extracellular mycocardial acidosis during acute ischemia, gastroesophageal reflux monitoring and seat analysis [48]. In one of the designs, a flexible IrO_x-based pH sensor on a polymeric substrate by low-cost, simpler sol–gel fabrication method was developed [48]. The fabricated sensor consists of three pairs of electrodes (both sensitive and reference electrodes) on a polyamide substrate. This flexible sensor array was rolled up with a radius of 1 cm and inserted into a long tube with a diameter of 2 cm. The connection lines and contact wires of this pH sensor were designed at the millimetre scale for miniaturisation and easy handling, as shown in figures 8.3(a) and (b) [48]. By using the IrO_x sensitive films, the sensor exhibits promising sensitivity, response time, stability, repeatability, reversibility and selectivity. The reversibility and repeatability in performances are given in figure 8.3(c) [48]. There are various methods implemented for the development of potentiometric flexible pH sensors, including sputtering [49, 50], sol–gel deposition [48], anodic electrodeposition [51], etc. Various deposition methods for flexible IrO_2-based pH sensors are given in [52].

The screen-printing approach was found to be a simple and low-cost method for flexible pH sensor fabrication. In one of the works, a disposable RuO_2-based potentiometric pH sensor by the screen-printing technique [53]. A standard saturated calomel electrode was used as a reference electrode. The prepared sensitive electrode shows faster response in acidic solution than alkaline solution, and good sensitivity (51 mV/pH) [53]. The potentiometric performance illustrated that the RuO_2 sensitive electrode does not need pretreatment, has negligible interference from other ions and no hysteresis effect [53]. Printing allows the development of various designs of sensors, including interdigitated electrode (IDE) based

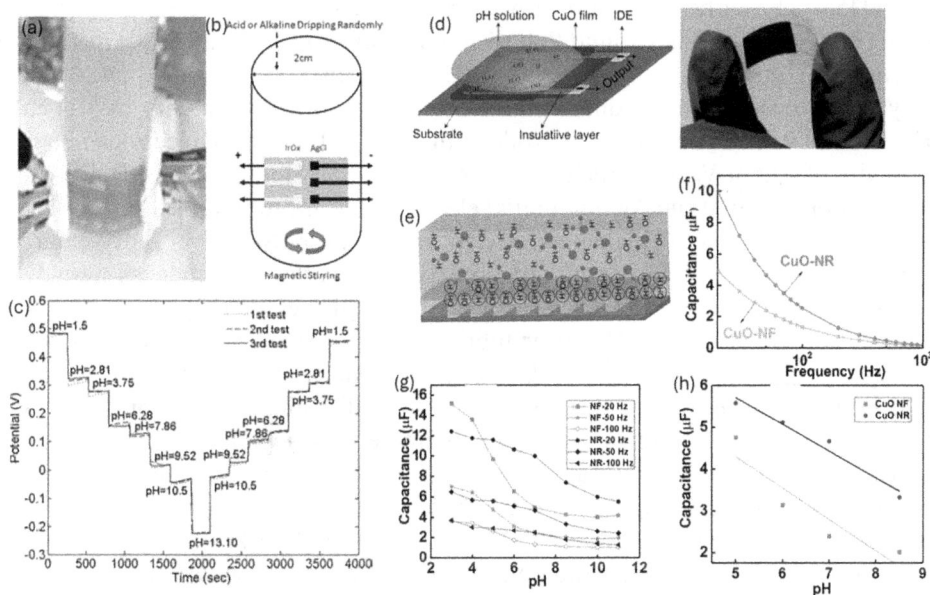

Figure 8.3. (a) Image of flexible IrO_2-based pH sensor, (b) the design of the sensor, (c) measured results for reversibility and repeatability experiments with pH (1.5–13.1) for three tests, reprinted from [48], copyright (2011), with permission from Elsevier, (d) screen printed CuO nanostructures on IDE configuration, (e) electrochemical double layer formation of CuO-based sensor, (f) variation of capacitance with frequency for sensors, (g) capacitance variation with pH values for different frequencies, (h) change in capacitance in the region of pH 5–8.5 for CuO nano rectangle and nanoflower-based sensors at 50 Hz, reprinted from [54], copyright (2018), with permission from Elsevier.

conductimetric pH sensor as shown in figure 8.3(d) [54]. For the sensor design, CuO nanoflowers and nano rectangles were prepared by the hydrothermal method, and based on the prepared powder, a screen printable paste was developed. Due to high crystallinity and low surface roughness (130 nm) of CuO nano rectangle compared with respect to CuO nanoflower (surface roughness—192 nm), the CuO nano rectangle-based sensors show excellent performance [54]. In these conductimetric or capacitive-based sensors, while varying the pH value of the solution, the H^+ or OH^- charged surface groups of the electrical double layer, as shown in figure 8.3(e), formed at the interface of CuO–electrolyte. The change in capacitance of the sensor for both electrodes is given in figure 8.3(f), and its variation of capacitance with pH values for different frequencies is given in figure 8.3(g) [54]. The CuO nano rectangle-based sensor exhibits a sensitivity of 0.64 μF/pH in the range pH 5–8.5, as shown in figure 9.3(h) [54].

For improving the accuracy and reliability of the flexible sensor, a data fusion and fault diagnosis measurement system was implemented based on LabVIEW (Laboratory Virtual Instrumentation Engineering Workbench), which removes the fault sensing window [55]. In this work, a flexible arrayed pH sensor based on a RuO_2 sensitive electrode and an Ag differential reference electrode were developed [55]. The RuO_2 was deposited on a PET substrate by RF sputtering and an Ag differential electrode was prepared on the same substrate by the screen-printing method [55]. The authors also used a Ag/AgCl reference electrode for comparing the sensing performance. They found that the data fusion and fault diagnosis method of measurement improves the sensitivity ~22% and linearity by 0.14. The flexible arrayed pH sensor with Ag differential electrode exhibits a sensitivity of 57.1 mV/pH, whereas a RuO_2 sensitive electrode with Ag/AgCl reference electrode shows a sensitivity of 53.6 mV/pH in the range of pH 1–13 [55]. Hence, from this work, the authors proposed that the Ag differential electrode was more suitable for flexible pH sensor fabrication due to higher sensitivity, and it can be stored in a dry environment. However, the pH sensor with an Ag differential electrode shows lower linearity than a Ag/AgCl reference electrode-based pH sensor [55].

Recent investigation in pH sensor fabrication for biomedical, clinical and food monitoring is concentrated towards the pH extended gate thin-film transistors (EGFET) sensors because of their full device flexibility [56, 57]. One of the designs of pH EGFET measurement is shown in figure 8.4(a) [56]. In which sensitive amorphous RuO_x was coated via the sol sol–gel method. Here, a flexible ITO-coated PET was used for the coating of RuO_x-based sensitive electrode in pH EGFET sensor [56]. The developed sensor exhibits a sensitivity of 65.11 mV/pH in the wide range of pH 2–12. The cyclic bending analysis of the sensors shows up to 500 times of bending a negligible variation in sensitivity and linearity of the sensor. However, due to the generation of deformation in the microstructure and cracking in the sensing film after 700 cycles of bending, a sudden drop in sensitivity was observed, as shown in figure 8.4(b) [56]. Maiolo *et al* [58] developed a fully flexible pH-EGTFT based on LTPS (low temperature polycrystalline silicon) TFTs (thin-film transistors). ZnO nanowalls were used as a sensing layer, and deposited on polymer substrates (PI and polyarylate (PAR). ZnO-based sensors exhibit a response close to

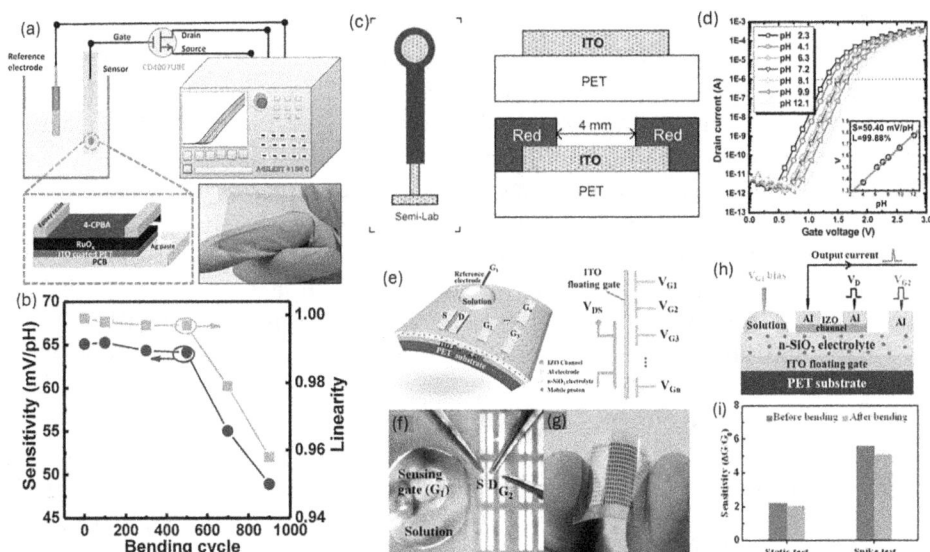

Figure 8.4. (a) Schematic representation of device configuration and measurement of flexible pH-EGFET based on RuOx sensing film, reprinted from [56], copyright (2019), with permission from Elsevier. (b) Variation of sensitivity factor and linearity of the EGFET with bending cycles, reprinted from [56], copyright (2019), with permission from Elsevier. (c) Schematic and cross-section of the ITO/PET electrode, reprinted from [59], copyright (2012), with permission from Elsevier. (d) I_{DS}–V_{GS} curves of 100 Ω $\square^{-1}$ ITO/PET-EGFET samples measured in buffer solutions in the pH range pH 2.3–12, reprinted from [59], copyright (2012), with permission from Elsevier. (e) Schematic representation of the flexible pH sensor based on an IZO neuromorphic transistor with multiple gate electrodes, with a capacitive network of the transistor. (f) Optical microscopic image of the measurement set up. (g) Array of the flexible sensor. (h) Schematic image of spike measurement and its mechanism. (i) Influence of bending on the sensitivity of the transistor for both quasi-static and single-spike sensing modes (after 1000 times). (e)–(i) reproduced from [60] CC BY 4.0

the ideal Nernstian response. With large surface-to-volume ratio and also the presence of surfactant defects and/or vacancies, the ZnO nanostructures improve the sensing performances [58]. In addition to the above properties, ZnO nanostructures show very good thermal and chemical stabilities, mechanical strength, non-toxicity and electrochemical activity. Due to these properties ZnO-based sensors have very good applications in the biosensor field [58]. Lue *et al* [59] proposed an ITO (indium tin oxide—100 Ω $\square^{-1}$) sensing layer deposited on flexible PET substrate function as a very good pH EGFET. A schematic representation of the sensor is shown in figure 8.4(c) [59]. To investigate the pH sensitivity of the ITO/PET-EGFET, drain-to-source current–gate-to-source voltage (I_{DS}–V_{GS}) curves were measured and are shown in figure 8.4(d) [59].

Liu *et al* [60] reported for the first time flexible electrolyte-gated neuron transistors with amorphous oxide (IZO—Indium Zinc Oxide) channel layers for biochemical pH sensing applications. Amorphous oxide-based transistors have significant importance in sensor fabrication due to their excellent electrical properties, low process temperature, high reliability and easy reproducibility [58, 60, 61]. Moreover, the electrolyte-gated transistors overcome the high operating voltage for portable systems. The major advantage of an electrolyte-gated transistor-based sensor is the strong electrical double layer (*edl*) modulation at the electrolyte/oxide interface that reduces the operation voltage of the sensor. Figure 8.4(e) shows the IZO-based neuromorphic transistor and its flexibility performance. The measurement system and flexible sensor array is shown in figures 8.4(f) and 8.4(g). It was found that the dual-gate synergic modulation mode improves the sensitivity (105 mV/pH) compared to the single-gate mode (37.5 mV/pH) operation of the sensor. It was observed that in single-spike dynamic sensing mode (figure 8.4(h)) the energy consumption is extremely low, and it improves the sensing performance and reproducibility [60]. In addition, the sensitivity reduction is less than 10% for both quasi-static and single-spike sensing modes after 1000 bending cycles, as shown in figure 8.4(i).

One of the great advantages of polymer-based pH sensors, as discussed in the previous chapter, is their applications in flexible electronics fabrication. However, polymers show limited chemical stability and low mechanical strength. Recently, significant research has been reported in the development of wearable pH sensors by using polymers or organic composite materials. One of the simple approaches considered for the sensor development is screen printing on a bandage or other fabric substrates. As an example, a bandage-based pH sensor was fabricated by screen printing on a Ag/AgCl reference and electropolymerized polyaniline (PANi) as a sensitive electrode [62]. The steps of fabrication are shown in figures 8.5(a) and (b) and represent the developed pH sensor [62]. The bendability of the smart bandage-based pH sensor is shown in figure 8.5(c) [62]. The pH bandage sensor exhibits excellent response to various pH solutions and the pH sensitivity close to the theoretical Nernstian response (59.2 mV/pH) as shown in figure 8.5(d) in the pH range 4.35–8 [62]. The sensor shows less interference to other ions, fast response time, good repeatability (as shown in figure 8.5(e)), reproducibility and lack of hysteresis effect [62]. The fabricated sensor shows a minimal impact on sensing performance with different bending cycles and application in human serum (due to its similarity with the chemical environment in the vicinity of a wound) [62]. For flexible polymer-based pH sensor fabrication, PANI as an ion-sensitive electrode was found of prime interest, including the design of a tattoo-based pH sensor, as shown in figure 8.5(f) [33]. The 'smiley face'-shaped pH sensor was fabricated by using Ag/AgCl as a reference electrode, on a temporary transfer tattoo paper. The fabricated pH sensor shows promising potentiometric performance for monitoring pH in the range of 3–7, which is the pH range of human sweat [33]. The key parameter of a wearable sensor for practical applications is its sensing performance under mechanical strain (bending or stretching). Figure 8.5(g) represents the different mechanical strain conditions of the tattoo pH sensor. The real-time measurement of the pH levels of human body skin reveals that the fabricated tattoo-based pH sensors are suitable for practical applications [33].

In addition to PANI alone, a mixed conducting polymer (PANI and PPY) was also used as a sensing layer [64]. The sensing performance of the fabricated interdigitated conductimetric flexible pH sensor reveals that the pH value is almost linearly related to the electrical conductivity in the pH range 4–7 [64]. In a low

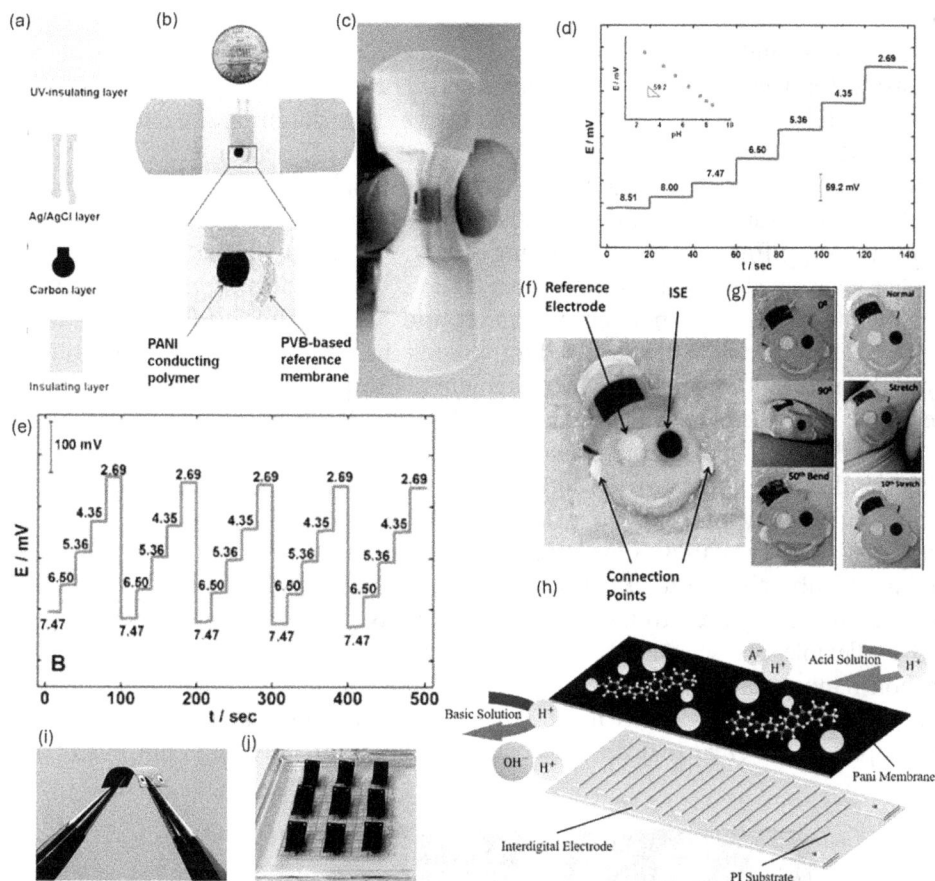

Figure 8.5. (a) Fabrication steps to create the pH-sensitive bandage [4]. (b) Image showing the actual bandage-based sensor. (c) Image of bending of the pH bandage sensor. (d) Potentiometric performance of pH bandage sensor from pH 8.51–2.69. (e) Repeatability of a pH bandage sensor within the pH 7.47–2.69 range. (a)–(e) [62] John Wiley & Sons. Copyright 2014 WILEY-VCH Verlag GmbH & Co. KGaA, Weinheim. (f) Image of 'smiley face'-shaped tattoo pH sensor. Panel (g) represents the bending and stretching performance of the sensor. (f), (g) reproduced from [33] with permission from the Royal Society of Chemistry. (h) Schematic representation of IDE-based pH sensor with PANI as sensitive film and its protonation and deprotonation. (i) Image of the flexible IDE-based pH sensor. (j) Photograph of the fabricated multiple sensor. (h)–(j) reproduced with permission from [63] CC-BY-NC 3.0.

temperature proton exchange membrane fuel cell (PEMFC) the acidic environment of the fluid in the fuel cell, corrosion etc, influence the performance of the fuel cell. So, for monitoring internal parameters of the fuel cell instantly, the flexible micro pH sensors like IDE structure have been useful [64]. In an IDE-based sensor, the variation in ionic concentration on the surface of the electrode changes in the sensor's resistance. One of the recent designs on the IDE-based sensor is given in figure 8.5(h) [63]. The sensor flexibility and possibility of mass production are shown in figures 8.5(i) and (j) [63]. In addition to this, for curved surface applications, one of the recent works, it was found that the electropolymerization of PANI on carbon cloth exhibits excellent pH sensitivity (60.9 mV/pH) [65].

Kaempgen and Roth [66] developed a new concept in the development of a substrate for flexible pH sensor fabrication. They introduced a thin-film carbon nanotube (CNT) network used as a conductive, transparent and flexible substrate for the deposition of PANI sensitive material for pH sensor fabrication. CNT networks form an electrical contact to the sensitive layer and allow both optical and electronic measurements simultaneously. This type of substrate overcomes the problems of stiffness, size and limiting substrates, and it opens a wide range of applications in electrochemical sensors and CNTs. In this work, a transparent polymer film (made of polyteraphthalate) and also a wire (steel wire coated by polyvinyl chloride (PVC)) were used for coating the CNT network as a substrate for optical and potentiometric pH measurements. The pH-sensitive PANI was deposited by electrochemical [66]. Figure 8.6(a) shows the fabricated transparent and flexible CNT/Pani pH sensors. Kaempgen and Roth observed that the PANI-coated film significantly improved the linear response (shown in figure 8.6(b)), response time and reproducibility of the pH sensor as compared to a pure CNT-based pH sensor. Moreover, Pani/CNT pH sensors exhibit lower drift and hysteresis effects and are highly selective to H^+. The sensor also shows high transmittance in the acid pH region as compared to the basic pH region, as shown in figure 8.6(c) [66].

For flexible pH sensor fabrication, in addition to metal oxides and polymer, carbon nanostructures have also attracted significant interest. Table 8.1 compares the performances of various flexible pH sensors and its materials used for their development. Li et al [67] developed a microfluidic pH sensing chip based on a

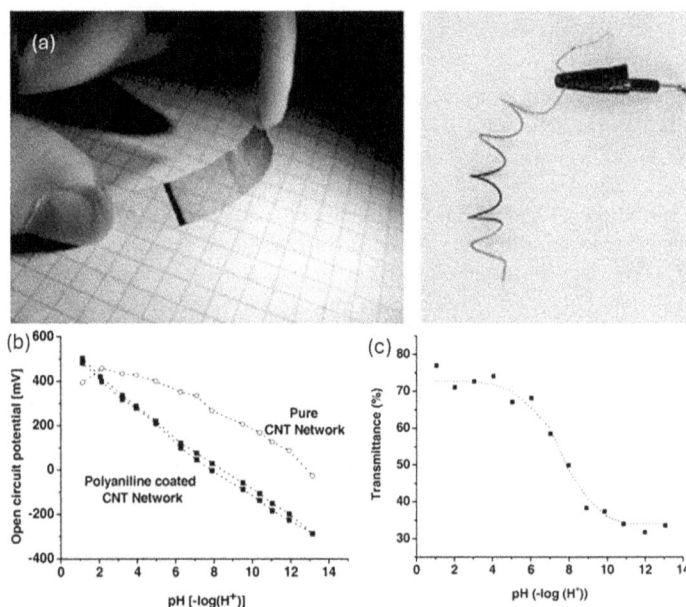

Figure 8.6. (a) Transparent and flexible CNT/PANI pH sensors. (b) The pH dependence of the open circuit potential of a CNT/PANI pH sensor. (c) Transmittance of thin film CNT/PANI at $k = 575$ nm after immersing in various buffer solutions. Reprinted from [66], copyright (2006), with permission from Elsevier.

Table 8.1. Comparison of the performance of various materials-based flexible and stretchable pH sensors.

Material	Fabrication	Substrate	pH range	Response time	Sensitivity (mV/pH)	Flexibility	References
IrO$_2$	Sputtering	PET	4–7	—	61 ± 1	—	[69]
IrO$_2$	Electrodeposition	PI	4–9	0.5 s	69.9 ± 2.2	—	[70]
IrO$_2$	Sol–gel	PI	1.5–12	0.9–2 s	51	Tested in a tube with a 1 cm curvature radius	[48]
CuO nano rectangle	Hydrothermal synthesis	PET	5–8.5	—	0.64 μF/pH	Tested in a tube with a 5 mm curvature radius	[54]
ZnO nanowalls	Low temperature polycrystalline silicon thin film transistor technology	PI	1–9	—	~59	—	[58]
IZO	Sputtering	PET	4–10	5 ms	105	Tested by bending around a cylinder with a 1.0 cm curvature radius	[60]
ITO	Radio frequency sputtering and a roll-to-roll process	PET	2–12	—	50.1	—	[59]
InGaZnO	Sputtering and thin film transistor technology	PI	3.3–11	—	51.2	Tested as a function of bending with up to 13 mm curvature radius	[71]
WO$_3$ nanoparticle	Electrodeposition	PI	5–9	23–28 s	−56.7 ± 1.3	—	[72]
PANi	Electrodeposition	a. PET b. PVC coated steel wire	1–13	A few seconds	58	—	[66]
PANi	Electrodeposition	PDMS	4–7	~60 s	—	Tested by mechanical friction and skin deformation	[73]

Material	Method	Substrate	pH range	Response time	Sensitivity (mV/pH)	Notes	Reference
PANi	Drop-casting	PET-coated palette paper	4–10	Rise time 12 s Fall time 36 s (pH 6–8)	50–58.2 (pH 2–12)	—	[74]
PANi	Electropolymerization	Commercial adhesive bandage	5.5–8	20 s	58.0 ± 0.3	Tested by flexing the sensor and then releasing the device to return to its unperturbed state (100 iterations)	[75]
PANi	Electropolymerization	Commercially available temporary transfer tattoo paper	3–7	25 s	52.8–59.6 (dependent on bending/ stretching conditions)	Tested using GORE-TEX under 50 bending (180°) and 40 stretching (10% in lateral extent) applications	[33]
PANi	Electrodeposition	Parylene C-coated newspaper	2–12	<10 s	58.2	Tested on a glass rod with respect to a bending radius of 7 mm	[76]
PANi	Dilute chemical polymerization	PET	3.9–10.1	12.8 s	62.4		[77, 78]
PAA-CNTs	Electropolymerization	Si-chips	2–12	3 s	54.5	—	[79]
SWCNT	Vacuum filtration method	PET	3–11	30 s	59.71	Tested by measuring resistivity upon hard bending	[37, 67]
G-PU	Printing	PDMS	5–9	8 s	11.13 ± 5.8	Tested by measuring resistance under 500 stretching cycles (30% strain), with the use of a stepper motor	[37]
G-PU	Printing	Cellulose-polyester blend cloth	6–9	5 s	47 ± 2	Demonstrated by hard crumpling Tested under 500 bending cycles at 11.40 mm bending radius	[32]

single-walled carbon nanotube (SWCNT) thin film. SWCNTs are a promising material for electrochemical sensors because they improve the potentiometric stability and exhibit good electrical, thermal, chemical and mechanical properties. SWCNTs were deposited on both glass and flexible (PET) substrates by the vacuum filtration method. For reference electrode fabrication, an Ag/AgCl paste was painted on one of the SWCNT electrodes. The fabrication steps of the microfluidic chip pH sensor are shown in figure 8.7(a), and the image of the flexible SWCNT-based pH sensor is shown in figure 8.7(b). The fabricated electrodes perform as a promising pH sensor (potentiometric performance shown in figure 8.7(c)) with an acceptable limit of selectivity coefficient towards other ions. The fabricated sensor is suitable for a flow analysis measurement system and applicable to the detection of metabolic processes in biological cells [67]. Due to the biocompatibility with various biological samples, paper-based analytical devices have significant applications in microfluidic devices, chemical and biosensors. Lei *et al* [68] observed that a CNT-based pH sensor on paper has been demonstrated to detect the pH value of an analyte solution. They observed that the resistance change decreased with an increase of pH value [68].

In addition to carbon nanostructures, graphite-polyurethane (G-PU) composite also performs as an excellent sensitive material for disposable pH sensor fabrication on textiles as a substrate [32]. Schematic representation and image of potentiometric

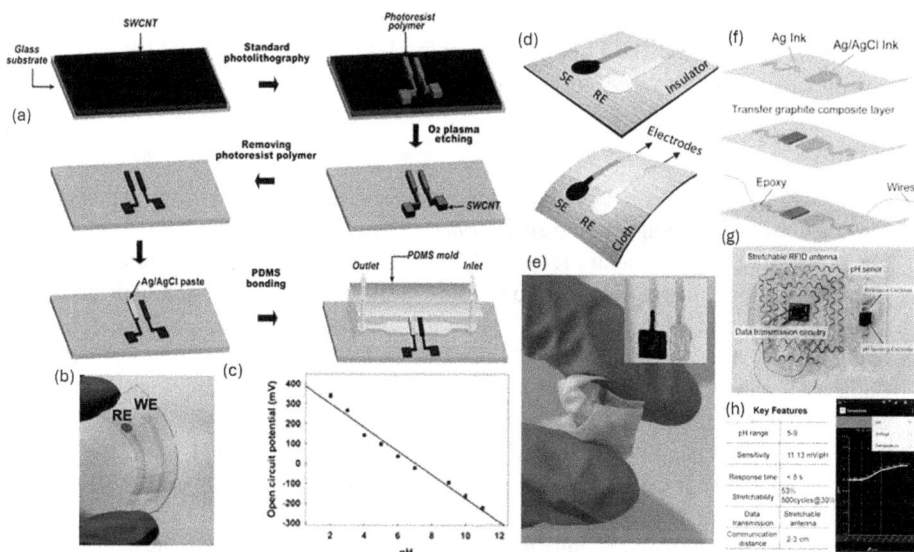

Figure 8.7. (a) Fabrication steps of microfluidic pH sensing chip based on SWCNTs. (b) Image of the flexible sensor. (c) Potentiometric performance of the sensor (a)–(c) reprinted from [67], copyright (2006), with permission from Elsevier. (d) Schematic of G-PU paste-based potentiometric pH sensor on textile. (e) Image of the flexible G-PU based pH sensor. (d), (e) reproduced from [32] CC BY 4.0. (f) Schematic of G-PU paste-based stretchable potentiometric pH sensor. (g) Image of the stretchable RFID antenna and pH sensors on the PDMS substrate. (h) Screenshot of smartphone App 'SenseAble' and photo of real-time pH monitoring system. (f)–(h) reprinted from [37], copyright (2018), with permission from Elsevier.

flexible pH sensor fabricated on textiles are shown in figures 8.7(d) and (e) [32]. The printing technology enables deposition of G-PU and a Ag/AgCl paste-based reference electrode. The G-PU pH-sensitive composite reacts with a pH solution, and an edl is formed at the interface between the electrode solutions [32]. The change in electrical properties such as impedance and capacitance of the edl depends on several parameters such as the electrode materials, the interface between electrode and electrolyte, the type of electrolyte, temperature, etc. Variations in the pH value of the solution lead to changes in the electrochemical properties of the sensitive electrode. The G-PU composite also enables design of a stretchable pH potentiometric pH sensor on a PDMS substrate (design shown in figure 8.7(f)) with a sensitivity of 11.13 ± 5.8 mV/pH with a maximum response time of 8 s [37]. The sensors shows 30% strain, the average strain experienced by human skin, for more than 500 cycles [37]. This stretchable pH sensor also found application as an online monitoring of sweat by integrating with a communication system (shown in figure 8.7(g)) and its display on a mobile phone through a custom-designed app, as shown in figure 8.7(h) [37].

8.4 Summary

The development of wearable and flexible pH sensors, embedded into different types of substrates by using different materials and methods of fabrication, is a growing area of research in electrochemical biosensor technology. Recently, flexible and stretchable pH sensors have received considerable attention for the continuous monitoring of human health. This chapter provides recent progress of the design and fabrication of various flexible and stretchable pH sensors.

References

[1] Li X, Zhang Z-Y and Li F 2025 Flexible electrochemical sensors based on nanomaterials: constructions, applications and prospects *Chem. Eng. J.* **504** 158101
[2] Markapudi P R, Beg M, Kadara R O, Paul F, Kerrouche A, See C H and Manjakkal L 2025 Nitrate pollution mapping for reservoirs using flexible sensors integrated with underwater robot *IEEE Internet Things J.* **12** 39172–80
[3] Manjakkal L, Mitra S, Petillot Y R, Shutler J, Scott E M, Willander M and Dahiya R 2021 Connected sensors, innovative sensor deployment, and intelligent data analysis for online water quality monitoring *IEEE Internet Things J.* **8** 13805–24
[4] Qin Y, Alam A U, Pan S, Howlader M M R, Ghosh R, Hu N-X, Jin H, Dong S, Chen C-H and Deen M J 2018 Integrated water quality monitoring system with pH, free chlorine, and temperature sensors *Sens. Actuators* B **255** 781–90
[5] Manjakkal L, Dervin S and Dahiya R 2020 Flexible potentiometric pH sensors for wearable systems *RSC Adv.* **10** 8594–617
[6] Neethirajan S 2017 Recent advances in wearable sensors for animal health management *Sens. Bio-Sen. Res.* **12** 15–29
[7] Yoon J, Cho H-Y, Shin M, Choi H K, Lee T and Choi J-W 2020 Flexible electrochemical biosensors for healthcare monitoring *J. Mater. Chem.* B **8** 7303–18
[8] Yang A and Yan F 2021 Flexible electrochemical biosensors for health monitoring *ACS Appl. Electron. Mater.* **3** 53–67

[9] Qin M, Guo H, Dai Z, Yan X and Ning X 2019 Advances in flexible and wearable pH sensors for wound healing monitoring *J. Semicond.* **40** 111607

[10] Tang Y *et al* 2022 Recent advances in wearable potentiometric pH sensors *Membranes* **12** 504

[11] NajafiKhoshnoo S, Kim T, Tavares-Negrete J A, Pei X, Das P, Lee S W, Rajendran J and Esfandyarpour R 2023 A 3D nanomaterials-printed wearable, battery-free, biocompatible, flexible, and wireless pH sensor system for real-time health monitoring *Adv. Mater. Technol.* **8** 2201655

[12] Yoon J H, Kim S-M, Park H J, Kim Y K, Oh D X, Cho H-W, Lee K G, Hwang S Y, Park J and Choi B G 2020 Highly self-healable and flexible cable-type pH sensors for real-time monitoring of human fluids *Biosens. Bioelectron.* **150** 111946

[13] Mu B, Cao G, Zhang L, Zou Y and Xiao X 2021 Flexible wireless pH sensor system for fish monitoring *Sens. Bio-Sens. Res.* **34** 100465

[14] Mu B, Dong Y, Qian J, Wang M, Yang Y, Nikitina M A, Zhang L and Xiao X 2022 Hydrogel coating flexible pH sensor system for fish spoilage monitoring *Mater. Today Chem.* **26** 101183

[15] Alam A U, Qin Y, Nambiar S, Yeow J T, Howlader M M, Hu N-X and Deen M J 2018 Polymers and organic materials-based pH sensors for healthcare applications *Prog. Mater. Sci.* **96** 174–216

[16] Khan S, Ali S and Bermak A 2019 Recent developments in printing flexible and wearable sensing electronics for healthcare applications *Sensors* **19** 1230

[17] Chen L Y, Tee B C-K, Chortos A L, Schwartz G, Tse V, Lipomi D J, Wong H-S P, McConnell M V and Bao Z 2014 Continuous wireless pressure monitoring and mapping with ultra-small passive sensors for health monitoring and critical care *Nat. Commun.* **5** 5028

[18] Yokota R, Yamamoto S, Yano S, Sawaguchi T, Hasegawa M, Yamaguchi H, Ozawa H and Sato R 2001 Molecular design of heat resistant polyimides having excellent processability and high glass transition temperature *High Perform. Polym.* **13** S61–72

[19] Liaw D-J, Hsu P-N, Chen W-H and Lin S-L 2002 High glass transitions of new polyamides, polyimides, and poly (amide-imide) s containing a triphenylamine group: synthesis and characterization *Macromolecules* **35** 4669–76

[20] Sekitani T, Zschieschang U, Klauk H and Someya T 2010 Flexible organic transistors and circuits with extreme bending stability *Nat. Mater.* **9** 1015

[21] Kaltenbrunner M, Sekitani T, Reeder J, Yokota T, Kuribara K, Tokuhara T, Drack M, Schwödiauer R, Graz I and Bauer-Gogonea S 2013 An ultra-lightweight design for imperceptible plastic electronics *Nature* **499** 458

[22] Nomura K, Ohta H, Takagi A, Kamiya T, Hirano M and Hosono H 2004 Room-temperature fabrication of transparent flexible thin-film transistors using amorphous oxide semiconductors *Nature* **432** 488

[23] Kanao K, Harada S, Yamamoto Y, Honda W, Arie T, Akita S and Takei K 2015 Highly selective flexible tactile strain and temperature sensors against substrate bending for an artificial skin *RSC Adv.* **5** 30170–4

[24] Nie B, Li R, Cao J, Brandt J D and Pan T 2015 Flexible transparent iontronic film for interfacial capacitive pressure sensing *Adv. Mater.* **27** 6055–62

[25] Luo N, Dai W, Li C, Zhou Z, Lu L, Poon C C, Chen S C, Zhang Y and Zhao N 2016 Flexible piezoresistive sensor patch enabling ultralow power cuffless blood pressure measurement *Adv. Funct. Mater.* **26** 1178–87

[26] Cai L, Li J, Luan P, Dong H, Zhao D, Zhang Q, Zhang X, Tu M, Zeng Q and Zhou W 2012 Highly transparent and conductive stretchable conductors based on hierarchical reticulate single-walled carbon nanotube architecture *Adv. Funct. Mater.* **22** 5238–44

[27] Hu S, Ren X, Bachman M, Sims C E, Li G and Allbritton N 2002 Surface modification of poly (dimethylsiloxane) microfluidic devices by ultraviolet polymer grafting *Anal. Chem.* **74** 4117–23

[28] Yu C, Masarapu C, Rong J, Wei B and Jiang H 2009 Stretchable supercapacitors based on buckled single-walled carbon-nanotube macrofilms *Adv. Mater.* **21** 4793–7

[29] Barbier V, Tatoulian M, Li H, Arefi-Khonsari F, Ajdari A and Tabeling P 2006 Stable modification of PDMS surface properties by plasma polymerization: application to the formation of double emulsions in microfluidic systems *Langmuir* **22** 5230–2

[30] Guinovart T, Parrilla M, Crespo G A, Rius F X and Andrade F J 2013 Potentiometric sensors using cotton yarns, carbon nanotubes and polymeric membranes *Analyst* **138** 5208–15

[31] Smith R E, Totti S, Velliou E, Campagnolo P, Hingley-Wilson S M, Ward N I, Varcoe J R and Crean C 2019 Development of a novel highly conductive and flexible cotton yarn for wearable pH sensor technology *Sens. Actuators* B **287** 338–45

[32] Manjakkal L, Dang W, Yogeswaran N and Dahiya R 2019 Textile-based potentiometric electrochemical pH sensor for wearable applications *Biosensors* **9** 14

[33] Bandodkar A J, Hung V W, Jia W, Valdés-Ramírez G, Windmiller J R, Martinez A G, Ramírez J, Chan G, Kerman K and Wang J 2013 Tattoo-based potentiometric ion-selective sensors for epidermal pH monitoring *Analyst* **138** 123–8

[34] Bandodkar A J, Jia W and Wang J 2015 Tattoo-based wearable electrochemical devices: a review *Electroanalysis* **27** 562–72

[35] Shafiee H, Asghar W, Inci F, Yuksekkaya M, Jahangir M, Zhang M H, Durmus N G, Gurkan U A, Kuritzkes D R and Demirci U 2015 Paper and flexible substrates as materials for biosensing platforms to detect multiple biotargets *Sci. Rep.* **5** 8719

[36] Li R-Z, Hu A, Zhang T and Oakes K D 2014 Direct writing on paper of foldable capacitive touch pads with silver nanowire inks *ACS Appl. Mater. Interfaces* **6** 21721–9

[37] Dang W, Manjakkal L, Navaraj W T, Lorenzelli L, Vinciguerra V and Dahiya R 2018 Stretchable wireless system for sweat pH monitoring *Biosens. Bioelectron.* **107** 192–202

[38] Gao W *et al* 2016 Fully integrated wearable sensor arrays for multiplexed *in situ* perspiration analysis *Nature* **529** 509

[39] Yu Y, Nyein H Y Y, Gao W and Javey A 2020 Flexible electrochemical bioelectronics: the rise of *in situ* bioanalysis *Adv. Mater.* **32** e1902083

[40] Zhao J *et al* 2019 A fully integrated and self-powered smartwatch for continuous sweat glucose monitoring *ACS Sens.* **4** 1925–33

[41] Kim J, Campbell A S, de Ávila B E-F and Wang J 2019 Wearable biosensors for healthcare monitoring *Nat. Biotechnol.* **37** 389–406

[42] Vargas E, Teymourian H, Tehrani F, Eksin E, Sánchez-Tirado E, Warren P, Erdem A, Dassau E and Wang J 2019 Enzymatic/immunoassay dual-biomarker sensing chip: towards decentralized insulin/glucose detection *Angew. Chem. Int. Ed.* **58** 6376–9

[43] Kim J, Jeerapan I, Sempionatto J R, Barfidokht A, Mishra R K, Campbell A S, Hubble L J and Wang J 2018 Wearable bioelectronics: enzyme-based body-worn electronic devices *Acc. Chem. Res.* **51** 2820–8

[44] Kim J, Sempionatto J R, Imani S, Hartel M C, Barfidokht A, Tang G, Campbell A S, Mercier P P and Wang J 2018 Simultaneous monitoring of sweat and interstitial fluid using a single wearable biosensor platform *Adv. Sci.* **5** 1800880

[45] Tang Y, Zhong L, Wang W, He Y, Han T, Xu L, Mo X, Liu Z, Ma Y and Bao Y 2022 Recent advances in wearable potentiometric pH sensors *Membranes* **12** 504

[46] Chen Y-M, Chung T-W, Wu P-W and Chen P-C 2017 A cost-effective fabrication of iridium oxide films as biocompatible electrostimulation electrodes for neural interface applications *J. Alloys Compd.* **692** 339–45

[47] Zamora M L, Dominguez J M, Trujillo R M, Goy C B, Sánchez M A and Madrid R E 2018 Potentiometric textile-based pH sensor *Sens. Actuators* B **260** 601–8

[48] Huang W-D, Cao H, Deb S, Chiao M and Chiao J-C 2011 A flexible pH sensor based on the iridium oxide sensing film *Sens. Actuators* A **169** 1–11

[49] Maurya D K, Sardarinejad A and Alameh K 2014 Recent developments in R.F. Magnetron sputtered thin films for pH sensing applications—an overview *Coatings* **4** 756–71

[50] Chou J C, Yan S J, Liao Y H, Lai C H, Chen J S, Chen H Y, Tseng T W and Wu T Y 2018 Characterization of flexible arrayed pH sensor based on nickel oxide films *IEEE Sens. J.* **18** 605–12

[51] Prats-Alfonso E, Abad L, Casañ-Pastor N, Gonzalo-Ruiz J and Baldrich E 2013 Iridium oxide pH sensor for biomedical applications. Case urea–urease in real urine samples *Biosens. Bioelectron.* **39** 163–9

[52] Marsh P, Manjakkal L, Yang X, Huerta M, Le T, Thiel L, Chiao J C, Cao H and Dahiya R 2020 Flexible iridium oxide based pH sensor integrated with inductively coupled wireless transmission system for wearable applications *IEEE Sens. J.* **20** 5130–8

[53] Koncki R and Mascini M 1997 Screen-printed ruthenium dioxide electrodes for pH measurements *Anal. Chim. Acta* **351** 143–9

[54] Manjakkal L, Sakthivel B, Gopalakrishnan N and Dahiya R 2018 Printed flexible electrochemical pH sensors based on CuO nanorods *Sens. Actuators* B **263** 50–8

[55] Chou J-C, Lin C-Y, Liao Y-H, Chen J-T, Tsai Y-L, Chen J-L and Chou H-T 2013 Data fusion and fault diagnosis for flexible arrayed pH sensor measurement system based on LabVIEW *IEEE Sens. J.* **14** 1405–11

[56] Singh K, Lou B-S, Her J-L, Pang S-T and Pan T-M 2019 Super Nernstian pH response and enzyme-free detection of glucose using sol–gel derived RuOx on PET flexible-based extended-gate field-effect transistor *Sens. Actuators* B **298** 126837

[57] Pan T-M, Lin L-A, Ding H-Y, Her J-L and Pang S-T 2024 A simple and highly sensitive flexible sensor with extended-gate field-effect transistor for epinephrine detection utilizing InZnSnO sensing films *Talanta* **275** 126178

[58] Maiolo L, Mirabella S, Maita F, Alberti A, Minotti A, Strano V, Pecora A, Shacham-Diamand Y and Fortunato G 2014 Flexible pH sensors based on polysilicon thin film transistors and ZnO nanowalls *Appl. Phys. Lett.* **105** 093501

[59] Lue C-E, Wang I S, Huang C-H, Shiao Y-T, Wang H-C, Yang C-M, Hsu S-H, Chang C-Y, Wang W and Lai C-S 2012 pH sensing reliability of flexible ITO/PET electrodes on EGFETs prepared by a roll-to-roll process *Microelectron. Reliab.* **52** 1651–4

[60] Liu N, Zhu L Q, Feng P, Wan C J, Liu Y H, Shi Y and Wan Q 2015 Flexible sensory platform based on oxide-based neuromorphic transistors *Sci. Rep.* **5** 18082

[61] Fortunato E, Barquinha P and Martins R 2012 Oxide semiconductor thin-film transistors: a review of recent advances *Adv. Mater.* **24** 2945–86

[62] Guinovart T, Valdés-Ramírez G, Windmiller J R, Andrade F J and Wang J 2014 Bandage-based wearable potentiometric sensor for monitoring wound pH *Electroanalysis* **26** 1345–53

[63] Li Y, Mao Y, Xiao C, Xu X and Li X 2020 Flexible pH sensor based on a conductive PANI membrane for pH monitoring *RSC Adv.* **10** 21–8

[64] Lee C-Y, Chuang S-M, Lee S-J and Chiu C-Y 2016 Fabrication of flexible micro pH sensor for use in proton exchange membrane fuel cell *Int. J. Electrochem. Sci.* **11** 2263–8

[65] Hossain M S, Padmanathan N, Badal M M R, Razeeb K M and Jamal M 2024 Highly sensitive potentiometric pH sensor based on polyaniline modified carbon fiber cloth for food and pharmaceutical applications *ACS Omega.* **9** 40122–33

[66] Kaempgen M and Roth S 2006 Transparent and flexible carbon nanotube/polyaniline pH sensors *J. Electroanal. Chem.* **586** 72–6

[67] Li C A, Han K N, Pham X-H and Seong G H 2014 A single-walled carbon nanotube thin film-based pH-sensing microfluidic chip *Analyst* **139** 2011–5

[68] Lei K F, Lee K-F and Yang S-I 2012 Fabrication of carbon nanotube-based pH sensor for paper-based microfluidics *Microelectron. Eng.* **100** 1–5

[69] Nie C, Frijns A, Zevenbergen M and den Toonder J 2016 An integrated flex-microfluidic-Si chip device towards sweat sensing applications *Sens. Actuators* B **227** 427–37

[70] Chung H-J *et al* 2014 Stretchable, multiplexed pH sensors with demonstrations on rabbit and human hearts undergoing ischemia *Adv. Healthcare Mater.* **3** 59–68

[71] Nakata S, Arie T, Akita S and Takei K 2017 Wearable, flexible, and multifunctional healthcare device with an ISFET chemical sensor for simultaneous sweat pH and skin temperature monitoring *ACS Sens.* **2** 443–8

[72] Santos L *et al* 2014 WO₃ nanoparticle-based conformable pH sensor *ACS Appl. Mater. Interfaces* **6** 12226–34

[73] Lee H, Song C, Hong Y S, Kim M S, Cho H R, Kang T, Shin K, Choi S H, Hyeon T and Kim D-H 2017 Wearable/disposable sweat-based glucose monitoring device with multistage transdermal drug delivery module *Sci. Adv.* **3** e1601314

[74] Rahimi R, Ochoa M, Parupudi T, Zhao X, Yazdi I K, Dokmeci M R, Tamayol A, Khademhosseini A and Ziaie B 2016 A low-cost flexible pH sensor array for wound assessment *Sensors Actuators* B **229** 609–17

[75] Guinovart T, Valdés-Ramírez G, Windmiller J R, Andrade F J and Wang J 2014 Bandage-based wearable potentiometric sensor for monitoring wound pH *Electroanalysis* **26** 1345–53

[76] Yoon J H, Kim K H, Bae N H, Sim G S, Oh Y-J, Lee S J, Lee T J, Lee K G and Choi B G 2017 Fabrication of newspaper-based potentiometric platforms for flexible and disposable ion sensors *J. Colloid Interface Sci.* **508** 167–73

[77] Park H J, Yoon J H, Lee K G and Choi B G 2019 Potentiometric performance of flexible pH sensor based on polyaniline nanofiber arrays *Nano Converg.* **6** 9

[78] Dahiya R, Yogeswaran N, Liu F, Manjakkal L, Burdet E, Hayward V and Jörntell H 2019 Large-area soft e-skin: the challenges beyond sensor designs *Proc. IEEE* **107** 2016–33

[79] Gou P *et al* 2014 Carbon nanotube chemiresistor for wireless pH sensing *Sci. Rep.* **4** 4468

IOP Publishing

Advanced Electrochemical pH Sensing Technologies
Scientific fundamentals and applications
Libu Manjakkal

Chapter 9

Application of advanced electrochemical pH sensors and technological growth

9.1 Introduction

Miniaturised electrochemical pH sensors based on solid-state materials would be of great importance for use in biomedical, clinical diagnosis, environmental and food processing applications. Solid-state electrochemical pH sensors offer faster response, wider pH sensing range, excellent sensitivity, easier integration with microelectronic components for online application, wearable/flexible structural compatibility, better biocompatibility, integration on different substrates including clothes, paper, skin, etc, and lower cost of fabrication [1–3]. As compared to other analytical and optical methods, electrochemical pH sensing offers many advantages, as summarised in figure 9.1(a) [4]. The development of wearable and flexible pH sensors, embedded in different types of substrates by using different materials and methods of fabrication, is a growing area of research in electrochemical biosensor technology. For this, many materials such as metal oxide (MOx), carbon-based materials and organic semiconductors were employed for the flexible pH sensors. From the above chapters, it was found that the nanomaterials-based electrodes have large surface-to-volume ratios. These nanostructures enhance charge transfer abilities and enable improvement of the pH sensitivity. The advantages and weaknesses of the bulk, micro- and nanostructured materials are given in figure 9.1(b) [4]. From the detailed evaluation of the materials, the authors observed that various pH sensors with improved pH sensitivity were reported for online or real-time monitoring applications. A summary of different applications of the pH sensor is provided in figure 9.2. In this chapter we discuss the recent implementation of pH sensors in wearables, environmental and food quality monitoring applications.

doi:10.1088/978-0-7503-6079-1ch9
9-1

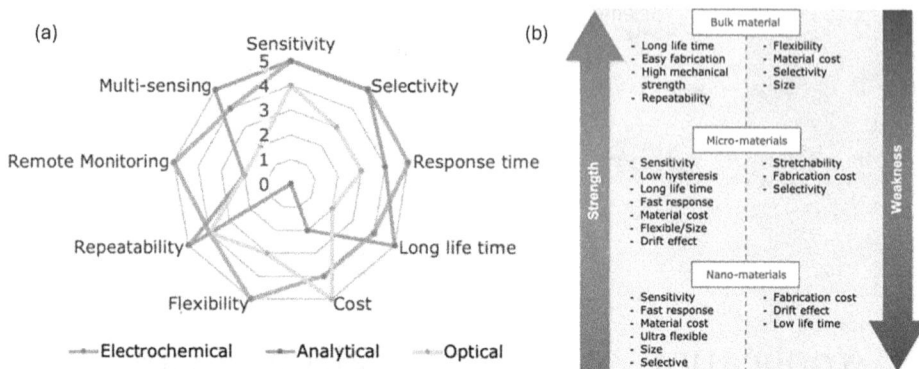

Figure 9.1. (a) Comparison of electrochemical, analytical, and optical-based sensors. (b) Comparison of strength and weakness factors of materials for sensor fabrication. Reproduced from [4] CC BY 4.0.

9.2 Wearable and biomedical applications of pH sensors

With improved healthcare in the last few decades, people are living longer, but with multiple health conditions, including chronic diseases. The daunting challenge today is how to care for an increasing number of individuals with complex medical conditions. There is a need to develop new solutions, such as sensors and wearable systems, that detect signs of chronic diseases early on and potentially enable self-health management [18–20]. The development of electrochemical biosensors that can be attached to the body or incorporated into clothing items has opened numerous possibilities for *in vivo* monitoring of patients over extended periods [21–23]. Wearable devices, including physical sensors, chemical sensors, optical sensors, and energy devices, can reveal crucial information regarding a wearer's health [24–27]. The sensing system is attached to body parts, textiles, contact lenses, etc, realising the application of wearable devices for online monitoring. The flexible and stretchable physical sensors fabricated based on microstructured and nano-structured materials can measure physical parameters such as pressure, temperature and strain. These physical sensors have been developed into wearable sensors that are in contact with the surface of organs or skin, aiding in disease diagnosis, human activity monitoring, and healthcare [28–36]. The major research works have been concentrating towards the application of physical sensors in the development of electronic skin (e-skin) and human–machine interfaces [37–41].

Chronic diseases such as diabetes, cancer, cardiovascular diseases and mental health disorders are the leading causes of death and disability worldwide. Early detection of chronic disease symptoms avoids the risks of many diseases and costly treatments. A wide range of health-related biomarkers, including cortisol, glucose, urea, lactate, uric acid, Na^+, K^+, Ca^{2+}, NH^{4+}, Cl^-, peptides, neuropeptides and cytokines etc, need to be monitored for early detection of diseases [3]. Laboratory-based analytical devices are not useful for real-time field-based applications. Integration of electrochemical sensors on wearable or flexible substrates opens new avenues in healthcare systems. It avoids sampling processes and interferences of

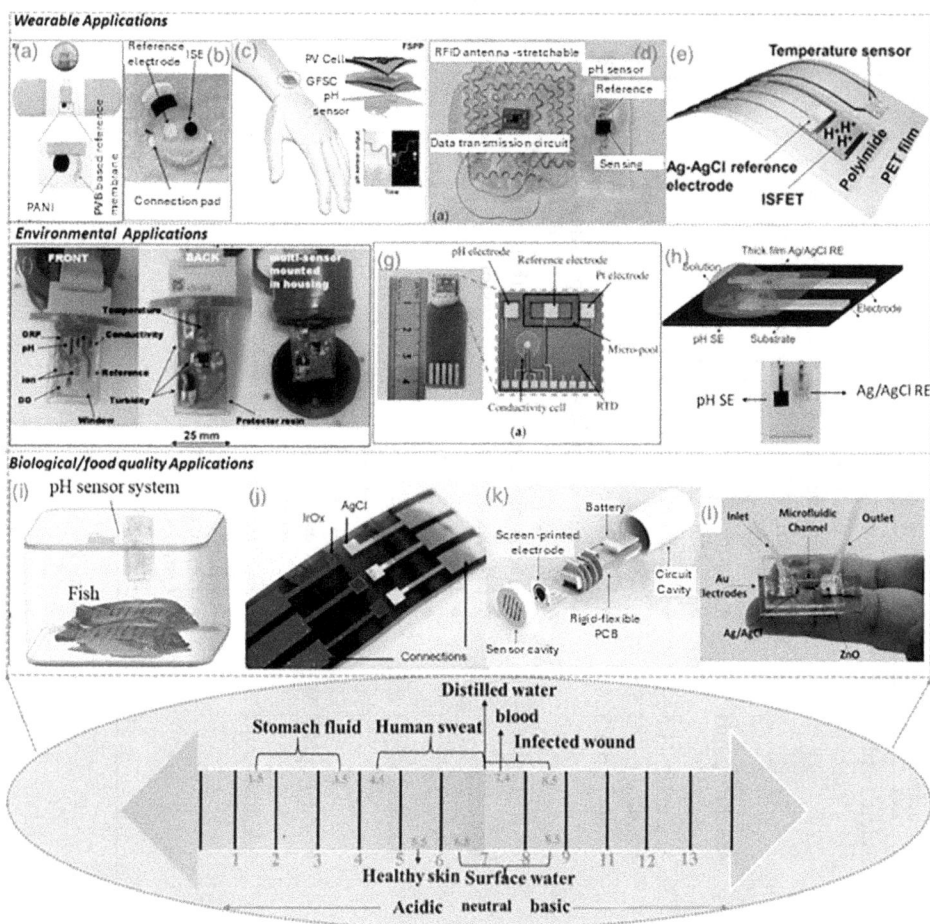

Figure 9.2. Application of pH sensors. (a) Bandage-based wearable pH sensor for wound monitoring [5] John Wiley & Sons. Copyright 2014 WILEY-VCH Verlag GmbH & Co. KGaA, Weinheim. (b) Tattoo-based ion SE epidermal pH monitoring. Reproduced from [6] with permission from the Royal Society of Chemistry. (c) Self-powered (integration of flexible solar cell and supercapacitor) flexible pH sensor. Reproduced from [7] CC BY 4.0. (d) Photo of stretchable wireless system (stretchable RFID and sensor) for sweat pH monitoring. Reprinted from [8], copyright (2018), with permission from Elsevier. (e) Schematic of a flexible pH and temperature sensor for wearable applications. Reprinted from [9], copyright (2017), with permission from Elsevier. (f) Online water pollution monitoring multisensors, including a pH sensor. Reprinted from [10], copyright (2012), with permission from Elsevier. (g) Photograph of the multi-parameter sensor chip (IrO$_2$-based pH sensor) for water pollution monitoring. Reproduced from [11] CC BY 4.0. (h) Schematic and image of a thick-film potentiometric pH sensor for water quality monitoring [12, 13], reproduced from [13] CC BY 4.0. (i) Wireless flexible pH sensor for fish spoilage monitoring. Reprinted from [14], copyright (2022), with permission from Elsevier. (j) Image of flexible IrO$_2$-based pH sensor. Reprinted from [15], copyright (2011), with permission from Elsevier. (k) Schematic view of the pH sensing capsule system based on iridium oxide for monitoring gastrointestinal health. Reprinted from [16], copyright (2021), with permission from Elsevier. (l) Photograph of the ZnO-based microfluidic pH sensor for quick recognition of circulating tumor cells in blood device. Reprinted with permission from [17], copyright (2017) American Chemical Society.

Table 9.1. The importance of pH monitoring in healthcare. Reproduced from [3] CC BY 3.0.

Body fluid	Function	Balanced pH	pH imbalance	Physiological status	References
Saliva	• Maintain healthy mouth • Protect teeth	6.2–7.6	Acidic (<pH 5.5) Alkaline (>pH 5.5)	• Demineralization and the breakdown of tooth enamel • Mineral deficiency (e.g. calcium and magnesium), often due to poor digestion • Chronic generalized periodontitis • Plaque formation • Chronic generalized gingivitis	[48–53]
Tears	• Prevent eye dryness	6.5–7.6	Acidic (<pH 5.5) Alkaline (>pH 5.5)	• Chemical damage	[54–57]
Urine	• Excrete waste fluid from the kidneys	4.5–8.0	Acidic (<pH 5.5) Alkaline (>pH 5.5)	• Metabolic syndrome • Diabetic ketoacidosis (a complication of diabetes) • Idiopathic uric acid nephrolithiasis (the process of forming a kidney stones) • Diarrhoea • Starvation • Kidney stones • Kidney-related disorders • Urinary tract infections (UTIs)	[58–62]
Sweat	• Control body temperature	4.5–7.0	Acidic (<pH 5.5) Alkaline (>pH 5.5)	• Acidosis • Excessive sweating • Electrolyte imbalance • Cystic fibrosis • Physical stress • Osteoporosis • Bone mineral loss	[63–69]

other analytes or contaminants while analysing a particular analyte in a body fluid. The flexibility of the devices enables rolling the sensors into a small diameter and allows them to be inserted into the body to measure chemical parameters of body fluids [42–44].

Among these wearable sensors, the flexible and stretchable electrochemical pH sensors are particularly important as the pH levels influence most chemical and biological reactions in materials, life and environmental sciences. The pH value can affect the activities of many physiological, biological and medical processes such as enzymatic reaction, tumor metastasis, wound healing, cell culture, etc [45–47]. Therefore, *in situ* assessment of pH of body fluids such as blood, urea, saliva and sweat could provide important information to early-stage detection of many diseases, as summarised in table 9.1 [3]. Moreover, the intracellular and extracellular pH an important determinant of many physiological processes [70]. The intracellular and extracellular acidosis influences myocardial performance [70]. The relative contribution of acidosis (acidosis means the increased acidity of blood) causes the development of ischemic heart diseases [70]. Conventional technologies for pH measurement based on glass electrode pH meters have many disadvantages, such as brittleness, large dimensions, periodical filling of reference buffer solution and calibrations, single-point measurement setup, slow response and limitations in high temperature and pressure in harsh environments. These disadvantages limit the implementation of this sensor for online or *in vivo* applications in biomedical, food processing and pollution monitoring.

Marzouk *et al* developed a pH electrode (first time reported anodic electro-deposited IrO_2) for measurement of extracellular mycocardial acidosis during acute ischemia [70]. Moreover, they also reported for the first time the simultaneous measurement of extrcellular pH, K^+ and lactate for cardiac physiology studies [70]. Grant *et al* reported the development of fiber optic and electrochemical based sensors to monitor brain tissue pH [71], in which the electrochemical pH sensor was fabricated on Kapton flexible substrate by using IrO_2 as sensitive membrane and Ag/AgCl as a reference electrode. The proposed sensor shows a Nernstian response of 57.9 ± 0.3 mV/pH for buffer solution and $57.8 \pm 1:5$ mV/pH for human blood in the range of pH 6.5–8.2 with a response time of less than 5 s [71]. Huang *et al* developed a flexible IrO_x-based pH sensor on a polymeric substrate by low-cost, simpler sol–gel fabrication method [15]. The fabricated sensor consists of three pairs of electrodes (both sensitive and reference electrodes) on a polyamide substrate. This flexible sensor array was rolled up with a curvature of radius of 1 cm and inserted into a long tube with a diameter of 2 cm. The connection lines and contact wires of this pH sensor were designed on a millimetre scale for miniaturisation and easy handling. By using the IrO_x sensitive films, the sensor exhibits promising sensitivity, response time, stability, repeatability, reversibility and selectivity. The fabricated sensors have great application in small confined tunnels, for example, *in vivo* reflux detection in the human esophagus [15]. For monitoring gastroesophageal reflux diseases it is necessary to monitor pH value for a long time in the stomach and esophagus [72, 73]. Based on this IrO_2 flexible pH sensor [15] Cao *et al* [72] reported a batteryless, wireless and implantable pH sensor along with an integrated impedance sensor for gastroesophageal reflux monitoring.

In this the impedance sensor was used to detect accurately the occurance of reflux episodes, both acidic and nonacidic. For the practical application of the sensor, live pigs under anesthesia were used. The fabricated miniaturized transponder does not required a battery and is small enough to be implanted on the wall of the oesophagus using endoscope [72]. Han et al [73] also reported a gastroesophageal tract pH sensor based on H^+-ion-sensitive field-effect transistor (ISFET). Nguyen et al [74] fabricated a pH sensor array consisting of 16 individual sensors (sensing electrode based on IrO_x and reference electrode based on Ag/AgCl) on a single polyamide flexible substrate by sol–gel method. The pH sensitivities of the sensors were in the range of 57.0–63.4 mV/ pH. For a flexible sensor it is very important to measure the substrate-deforming effect on the sensor performance. Nguyen et al observed that when the bending curvature radii were less than 8.4 cm, the change in sensitivities was less than 1 mV/pH. It was found that the sensors started to show the changes in sensing performance after bending the substrate with radius of 3 cm. So it is recommended to re-calibrate the sensor after each bending [74].

Monitoring of pH value of skin is important to detect pathogenesis of skin diseases like irritant contact dermatitis, acne vulgaris, atopic dermatitis, etc [8, 75–77]. The pH of skin is related to pH of sweat. Real-time sweat analysis is attractive for monitoring many factors including pH, Na^+, Cl^- ions; alchol and glucose level of sweat reflects their concentration in blood [78]. Microfluidic platform-based pH sensing devices operating on optical and electrochemical methods are successfully integrated for pH sensing of sweat. Curto et al [77] reported a disposable wearable chemical barcode for sweat pH sensing. Nie et al [75] introduced a flexible microfluidic device with an integrated silicon sensor chip for electrochemical pH monitoring of sweat. In this electrochemical sensor chip, the pH sensor is fabricated by using IrO_2 (prepared by sputtering method) as a sensitive electrode and Ag/AgCl RE. The fabricated sensor shows super-Nernstian response (61 mV/pH) in the pH range 2–10. When the chip pH sensor is attached to the skin, the paper absorbs sweat from the skin surface and fills the sensing cavity by capillarity [75].

The continuous monitoring of pH value of blood and brain tissue are very important for patients who have suffered stroke and traumatic brain injury. Grant et al [71] successfully fabricated fibre optic and electrochemical pH sensors to track the brain tissue pH. In this work, the authors developed the electrochemical pH sensor based on a thin multilayer coating of titanium, iridium and iridium oxide layers by sputtering onto a flexible Kapton substrate. The reference electrode (Ag/ AgCl) for this potentiometric pH sensor was also fabricated on Kapton substrate. The electrochemical pH sensor shows a Nernstian response of 57.9 ± 0.3 mV/pH in the range of pH value 6.8–8 with a response time of less than 5 s. The electro- chemical pH sensor can be inserted in the skull of traumatic head injury patients. For in vivo analysis, the sensing performance was carried out in Sprague-Dawley rats, and the sensor shows very little drift [71].

One of the greatest achievements in potentiometric pH sensors is the fabrication of novel wearable pH sensors for real-time monitoring of pH changes in a wound [2, 45, 79]. One of the proposed pH monitoring models is shown in figure 9.3(a). From this model, the comparison of estimated pyocyanin concentrations obtained from the

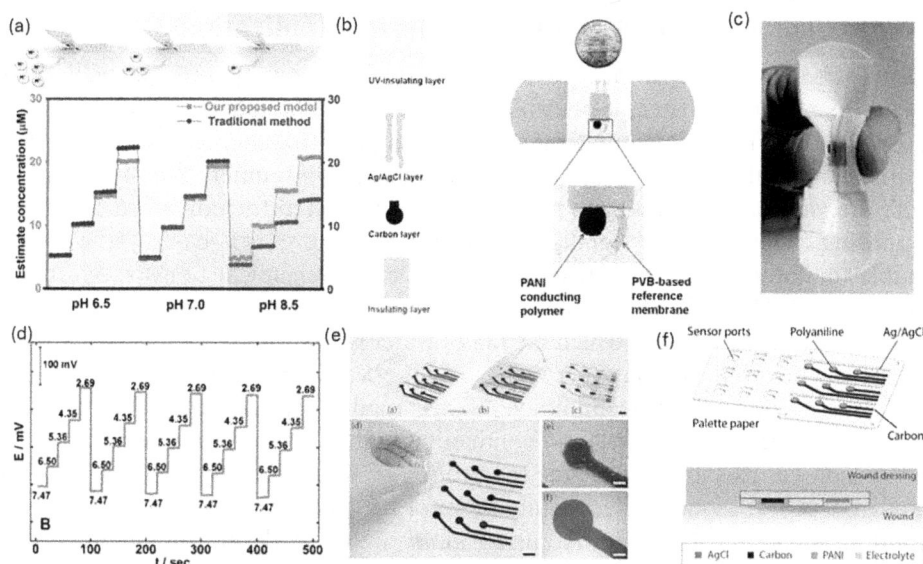

Figure 9.3. (a) Comparison of estimated pyocyanin concentrations obtained from the pH-correction model and the traditional calibration method at various pH levels of simulated wound fluids. Reproduced from [2] CC BY 4.0. (b) Fabrication steps to create the pH-sensitive bandage. (c) Image showing the actual bending of the pH bandage sensor. (d) Repeatability of a pH bandage sensor within the pH 7.47–2.69 range. (b)–(d) [5] John Wiley & Sons. Copyright 2014 WILEY-VCH Verlag GmbH & Co. KGaA, Weinheim. (e) Photograph of the flexible pH sensor array on a paper substrate with a close-up view of electrodes. (f) Schematic representation of the flexible pH sensor and cross-section of the sensor embedded into the wound dressing. (e), (f) reprinted from [80], copyright (2016), with permission from Elsevier.

pH-correction model and the traditional calibration method at various pH levels of simulated wound fluids are shown in figure 9.3(a) [2]. For healthy skin the pH value is approximately in the range of 5.5 but for infected wounds, pH value is in the range of 7–8.5. The alkaline nature of pH in the wound is due to the presence of bacterial colonies and enzymes. So, it's very important to the value of pH in wound assessment, especially for chronic wounds. However, it is difficult to use a bulky glass pH electrode for measuring or mapping pH levels of a wound. Considering the facts of simplicity of operation, low cost, compact size, real-time application, and to know the relation between pH value and wound healing, Guinovart *et al* developed a novel wearable pH sensor embedded in an adhesive bandage for real-time monitoring of pH changes in a wound, as shown in figures 9.3(b) and (c) [5]. The potentiometric pH sensing and repeatability of the bandage are shown in figure 9.3(d) [5]. Rahimi *et al* [80] presented an inexpensive flexible array of pH sensors fabricated on a palette paper substrate. The pH sensor is fabricated by using Ag/AgCl reference electrode and a sensitive electrode based on carbon coated with a conductive polymer, PANI. Figure 9.3(e) represents the photograph of the fabricated flexible pH sensor array and a close-up view of the electrodes [80]. The schematic view of the fabricated sensor and the cross-section view of the sensor array embedded into the wound dressing is shown in figure 9.3(f) [80]. The sensors show significant applications for integration with low-cost wound dressings [80].

9.3 Environmental food quality monitoring applications

The pH sensing in various water environments, soil and food quality, particularly in large areas and in real time, is complex, expensive and challenging for a number of reasons [4]. The real-time monitoring of various physical, chemical and biological parameters are also an important consideration for environmental monitoring [4]. There are various pH solid-state pH sensors implemented for online water quality monitoring. Due to the significant advantages, the RuO_2-based sensor was suitable for the implementation of many physical or chemical sensors. Martínez-Máñez *et al* [81] developed in thick-film technology a multisensor for measuring water quality parameters, in which the pH of water is measured by using RuO_2 paste as a sensing material. In this experiment, they fabricated the reference electrode Ag/AgCl by screen-printing technology. Water quality parameters, like dissolved oxygen, temperature, turbidity and conductivity, are also measured here [81]. This multisensor was executed for online monitoring of water pollution and to solve the drawbacks of traditional water quality monitoring. For online monitoring of water pollution, Zhuiykov [10] and Atkinson *et al* [82] also developed an integrated multisensor in which pH sensor was fabricated based on RuO_2 sensing electrode. In addition to this, a RuO_2-based pH sensor was also applied for measuring pH of some natural drinks, for example, Koncki and Mascini [83] developed a disposable pH sensor by screen-printing technique and studied its potential applications in pH measuring for some drinks, like Fanta, wine, milk, Coca-cola and orange juice, and the results are comparable with glass pH electrode. Liao and Chou [84] used RuO_2 thin film as a sensing layer of hydrogen in an ISFET. This pH sensor is also applied for measuring the pH of some drinks, like Coca-cola, lemon wine vinegar, water and milk. The measured pH values are in good agreement with the glass pH electrode [84].

In one of the recent works, the application of screen-printed Co_2O_3–RuO_2 composite electrodes-based pH sensor was directly implemented for the monitoring of pH value of natural environment for underwater and real-time measurement [85]. The proposed sensing system overcomes the drawbacks prevailing in field-to-laboratory water quality testing, such as time, logistics, technicalities, and alterations in sample quality [85]. The team carried out the pH sensing studies for different types of environmental water (lake, river and sea) in Poland, and developed a new underwater multisensor probe and analysed the quality of water in a lake at different depth levels in Germany. Figure 9.4(a) shows the block diagram and the probe developed for underwater monitoring [85]. Figure 9.4(b) demonstrates the proof-of-concept study of the sensor dipped in water, and the monitoring of water quality through the web server on a phone [85]. Figure 9.4(c) demonstrates the variation of temperature and pH value of water monitoring through a probe in different times [85]. The developed sensing technology has the potential to be further integrated with underwater robots for monitoring the quality in different depth levels and in different locations [4, 86]. The printed pH sensors, RuO_2-based pH sensors, also tested for soil monitoring for precision agriculture applications [87]. For the demonstration through the design of a printed circuit board (PCB) with

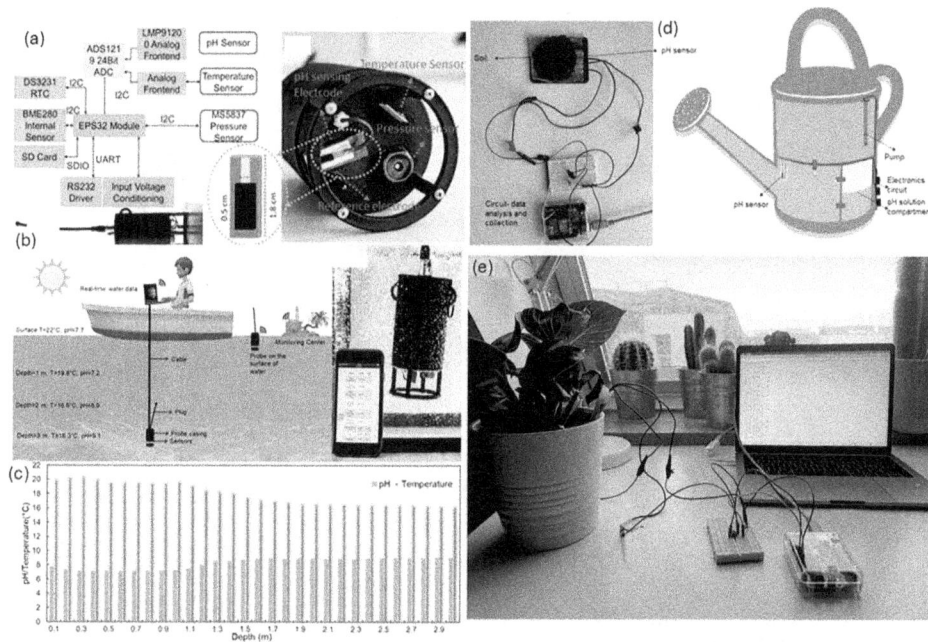

Figure 9.4. (a) Block diagram and developed pH sensing underwater probe. (b) Schematic representation of the measurement of pH and temperature using an underwater probe. Panel (c) shows changes in temperature and pH every 0.1 m from the surface, measured using the probe. (a)–(c) Reproduced from [85] with permission from the Royal Society of Chemistry. (d) Image and schematic representation of printed sensors with Arduino system and smart watering can for pH monitoring. (e) Image of the sensor used for pH monitoring of soil in precision agriculture. (d), (e) reprinted from [87], copyright (2023), with permission from Elsevier.

Arduino Uno board, we developed a prototype for online monitoring of soil and water pH and humidity. The sensing system is capable of monitoring pH value of soil and pouring water through the development of a new smart watering can, which can monitor water pH and adjust it through controlled addition of acidic/alkaline solution, as shown in figures 9.4(d) and (e) [87].

9.4 Integrated pH sensing system

Recent advances have witnessed the integration of wearable and flexible sensing platforms with low-powered electronics, suitable energy sources, and communication systems for health and environmental applications, as shown in figures 9.5(a) and (b) [1, 88]. For continuous monitoring, a long-lasting pH sensor and an ions monitoring system are also required. Along with the sensor, suitable power sources are a critical issue. There are multiple works reporting on the aspect of integrating solar cells with supercapacitors or batteries for powering the sensors and related components [1, 7, 8]. Current studies are progressing toward the power management and a suitable communication system for the sensing, as well as the data management using machine learning or the AI method.

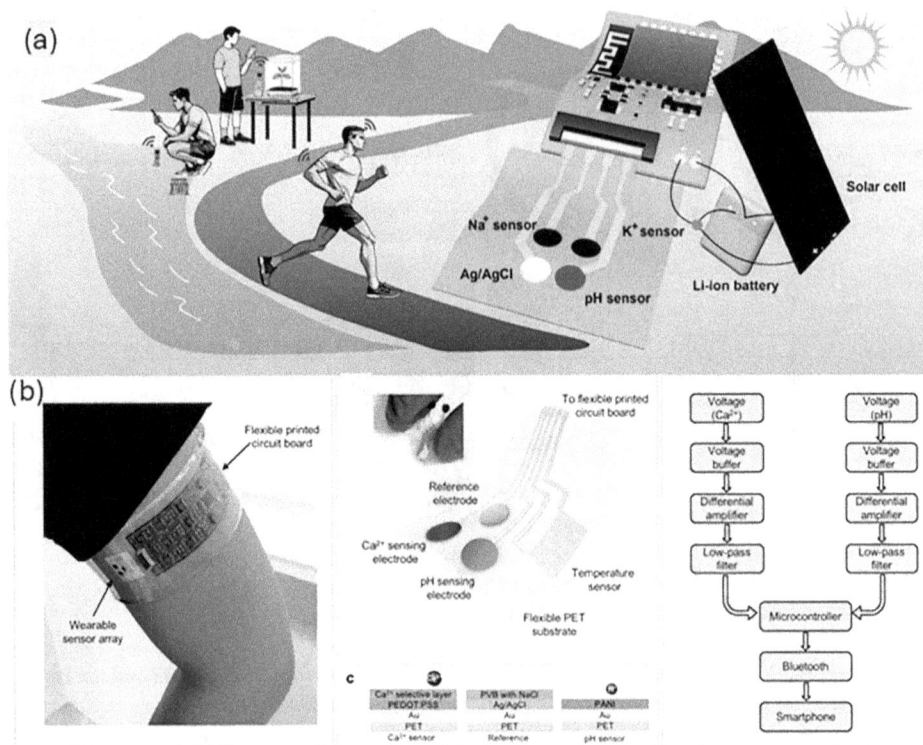

Figure 9.5. (a) Schematic illustration of the PES with capabilities of monitoring Na^+, K^+, and pH levels for complex outdoor environment applications. Reprinted with permission from [1], copyright (2025) American Chemical Society. (b) A fully integrated wearable multiplexed sensing system on a subject's arm, along with sensor design and schematic of a fully flexible system. Reprinted with permission from [88], copyright (2016) American Chemical Society.

9.5 Summary

Solid-state electrochemical pH sensors offer faster response, wider pH sensing range, excellent sensitivity, easier integration with microelectronic components for online application, wearable/flexible structural compatibility, better biocompatibility, integration on different substrates including clothes, paper, skin, etc, and lower cost of fabrication. In this chapter, a summary of different applications of the pH sensor is provided. This chapter discusses the recent implementation of pH sensors in wearables, environmental and food quality monitoring applications.

References

[1] Guan K, Wei R, Chen D, Jiang K, Kong X, Hua Q and Shen G 2025 Power-sustainable and portable electrochemical sensing platforms for complex outdoor environment applications *ACS Appl. Mater. Interfaces* **17** 3644–55

[2] Kaewpradub K, Veenuttranon K, Jantapaso H, Mittraparp-arthorn P and Jeerapan I 2024 A fully-printed wearable bandage-based electrochemical sensor with pH correction for wound infection monitoring *Nano-Micro Lett.* **17** 71

[3] Manjakkal L, Dervin S and Dahiya R 2020 Flexible potentiometric pH sensors for wearable systems *RSC Adv.* **10** 8594–617

[4] Manjakkal L, Mitra S, Petillot Y R, Shutler J, Scott E M, Willander M and Dahiya R 2021 Connected sensors, innovative sensor deployment, and intelligent data analysis for online water quality monitoring *IEEE Internet Things J.* **8** 13805–24

[5] Guinovart T, Valdés-Ramírez G, Windmiller J R, Andrade F J and Wang J 2014 Bandage-based wearable potentiometric sensor for monitoring wound pH *Electroanalysis* **26** 1345–53

[6] Bandodkar A J, Hung V W, Jia W, Valdés-Ramírez G, Windmiller J R, Martinez A G, Ramírez J, Chan G, Kerman K and Wang J 2013 Tattoo-based potentiometric ion-selective sensors for epidermal pH monitoring *Analyst* **138** 123–8

[7] Manjakkal L, Núñez C G, Dang W and Dahiya R 2018 Flexible self-charging super-capacitor based on graphene-Ag-3D graphene foam electrodes *Nano Energy* **51** 604–12

[8] Dang W, Manjakkal L, Navaraj W T, Lorenzelli L, Vinciguerra V and Dahiya R 2018 Stretchable wireless system for sweat pH monitoring *Biosens. Bioelectron.* **107** 192–202

[9] Nakata S, Arie T, Akita S and Takei K 2017 Wearable, flexible, and multifunctional healthcare device with an ISFET chemical sensor for simultaneous sweat pH and skin temperature monitoring *ACS Sens.* **2** 443–8

[10] Zhuiykov S 2012 Solid-state sensors monitoring parameters of water quality for the next generation of wireless sensor networks *Sens. Actuators* B **161** 1–20

[11] Zhou B, Bian C, Tong J and Xia S 2017 Fabrication of a miniature multi-parameter sensor chip for water quality assessment *Sensors* **17** 157

[12] Manjakkal L, Vilouras A and Dahiya R 2018 Screen printed thick film reference electrodes for electrochemical sensing *IEEE Sens. J.* **18** 7779–85

[13] Manjakkal L, Cvejin K, Kulawik J, Zaraska K, Szwagierczak D and Stojanovic G 2015 Sensing mechanism of RuO_2–SnO_2 thick film pH sensors studied by potentiometric method and electrochemical impedance spectroscopy *J. Electroanal. Chem.* **759** 82–90

[14] Mu B, Dong Y, Qian J, Wang M, Yang Y, Nikitina M A, Zhang L and Xiao X 2022 Hydrogel coating flexible pH sensor system for fish spoilage monitoring *Mater. Today Chem.* **26** 101183

[15] Huang W-D, Cao H, Deb S, Chiao M and Chiao J C 2011 A flexible pH sensor based on the iridium oxide sensing film *Sens. Actuators* A **169** 1–11

[16] Cheng C, Wu Y, Li X, An Z, Lu Y, Zhang F, Su B and Liu Q 2021 A wireless, ingestible pH sensing capsule system based on iridium oxide for monitoring gastrointestinal health *Sens. Actuators* B **349** 130781

[17] Mani G K, Morohoshi M, Yasoda Y, Yokoyama S, Kimura H and Tsuchiya K 2017 ZnO-based microfluidic pH sensor: a versatile approach for quick recognition of circulating tumor cells in blood *ACS Appl. Mater. Interfaces* **9** 5193–203

[18] Mujeeb-U-Rahman M, Nazari M H and Sencan M 2019 A novel semiconductor based wireless electrochemical sensing platform for chronic disease management *Biosens. Bioelectron.* **124** 66–74

[19] Subawickrama Mallika Widanaarachchige N R, Paul A, Banga I K, Bhide A, Muthukumar S and Prasad S 2025 Advancements in breathomics: special focus on electrochemical sensing and AI for chronic disease diagnosis and monitoring *ACS Omega* **10** 4187–96

[20] Min J, Sempionatto J R, Teymourian H, Wang J and Gao W 2021 Wearable electrochemical biosensors in North America *Biosens. Bioelectron.* **172** 112750

[21] Hatamie A, Angizi S, Kumar S, Pandey C M, Simchi A, Willander M and Malhotra B D 2020 Review—Textile based chemical and physical sensors for healthcare monitoring *J. Electrochem. Soc.* **167** 037546

[22] Khumngern S and Jeerapan I 2023 Advances in wearable electrochemical antibody-based sensors for cortisol sensing *Anal. Bioanal. Chem.* **415** 3863–77

[23] Wu Y, Mechael S S and Carmichael T B 2021 Wearable e-textiles using a textile-centric design approach *ACC. Chem. Res.* **54** 4051–64

[24] Manjakkal L, Yin L, Nathan A, Wang J and Dahiya R 2021 Energy autonomous sweat-based wearable systems *Adv. Mater.* **33** 2100899

[25] Manjakkal L, Pullanchiyodan A, Yogeswaran N, Hosseini E S and Dahiya R 2020 A Wearable supercapacitor based on conductive PEDOT:PSS-coated cloth and a sweat electrolyte *Adv. Mater.* **32** 1907254

[26] Hosseini E S, Bhattacharjee M, Manjakkal L and Dahiya R 2021 Healing and monitoring of chronic wounds: advances in wearable technologies ed A Godfrey and S Stuart *Digital Health* (New York: Academic) ch 6 pp 85–99

[27] Pullanchiyodan A, Manjakkal L, Ntagios M and Dahiya R 2021 MnOx-electrodeposited fabric-based stretchable supercapacitors with intrinsic strain sensing *ACS Appl. Mater. Interfaces* **13** 47581–92

[28] Heikenfeld J, Jajack A, Rogers J, Gutruf P, Tian L, Pan T, Li R, Khine M, Kim J and Wang J 2018 Wearable sensors: modalities, challenges, and prospects *Lab Chip* **18** 217–48

[29] Ates H C, Nguyen P Q, Gonzalez-Macia L, Morales-Narváez E, Güder F, Collins J J and Dincer C 2022 End-to-end design of wearable sensors *Nat. Rev. Mater.* **7** 887–907

[30] Azeem M, Shahid M, Masin I and Petru M 2025 Design and development of textile-based wearable sensors for real-time biomedical monitoring; a review *J. Text. Inst.* **116** 80–95

[31] Khan B, Riaz Z and Khoo B L 2024 Advancements in wearable sensors for cardiovascular disease detection for health monitoring *Mater. Sci. Eng.: R: Rep.* **159** 100804

[32] Arya S, Sharma A, Singh A, Ahmed A, Dubey A, Padha B, Khan S, Mahadeva R, Khosla A and Gupta V 2024 Energy and power requirements for wearable sensors *ECS Sens. Plus* **3** 022601

[33] Xue Z, Gai Y, Wu Y, liu Z and Li Z 2024 Wearable mechanical and electrochemical sensors for real-time health monitoring *Commun. Mater.* **5** 211

[34] Chen R, Jia X, Zhou H, Ren S, Han D, Li S and Gao Z 2024 Applications of MXenes in wearable sensing: advances, challenges, and prospects *Mater. Today* **75** 359–85

[35] Farzin M A, Naghib S M and Rabiee N 2024 Advancements in bio-inspired self-powered wireless sensors: materials, mechanisms, and biomedical applications *ACS Biomater. Sci. Eng.* **10** 1262–301

[36] Zhu J, Xia J, Li Y and Li Y 2025 Perspective on flexible organic solar cells for self-powered wearable applications *ACS Appl. Mater. Interfaces* **17** 5595–608

[37] Xiao X, Yin J, Xu J, Tat T and Chen J 2024 Advances in machine learning for wearable sensors *ACS Nano* **18** 22734–51

[38] Hu X, Wei Z, Sun Y, Zhang R and Chen C 2025 Multifunctional bilayer hydrogel electronic skin with thermochromic and electromagnetic interference shielding for wearable sensing applications *Compos. Commun.* **53** 102187

[39] Xu C *et al* 2024 A physicochemical-sensing electronic skin for stress response monitoring *Nat. Electron.* **7** 168–79

[40] Yang X, Chen W, Fan Q, Chen J, Chen Y, Lai F and Liu H 2024 Electronic skin for health monitoring systems: properties, functions, and applications *Adv. Mater.* **36** 2402542

[41] Gao B, Jiang J, Zhou S, Li J, Zhou Q and Li X 2024 Toward the next generation human–machine interaction: headworn wearable devices *Anal. Chem.* **96** 10477–87

[42] Yu A, Zhu M, Chen C, Li Y, Cui H, Liu S and Zhao Q 2024 Implantable flexible sensors for health monitoring *Adv. Healthc. Mater.* **13** 2302460

[43] Papani R, Li Y and Wang S 2024 Soft mechanical sensors for wearable and implantable applications *WIREs Nanomed. Nanobiotechnol.* **16** e1961

[44] Kong L *et al* 2024 Wireless technologies in flexible and wearable sensing: from materials design, system integration to applications *Adv. Mater.* **36** 2400333

[45] Li Y, Song S, Song J, Gong R and Abbas G 2025 Electrochemical pH sensor incorporated wearables for state-of-the-art wound care *ACS Sens.* **10** 1690–708

[46] Ghoneim M T, Nguyen A, Dereje N, Huang J, Moore G C, Murzynowski P J and Dagdeviren C 2019 Recent progress in electrochemical pH-sensing materials and configurations for biomedical applications *Chem. Rev.* **119** 5248–97

[47] Escobedo P *et al* 2021 Wireless wearable wristband for continuous sweat pH monitoring *Sens. Actuators* B **327** 128948

[48] Aguirre A, Testa-Weintraub L, Banderas J, Haraszthy G, Reddy M and Levine M 1993 Sialochemistry: a diagnostic tool? *Crit. Rev. Oral Biol. Med.* **4** 343–50

[49] Vasudev A, Kaushik A, Tomizawa Y, Norena N and Bhansali S 2013 An LTCC-based microfluidic system for label-free, electrochemical detection of cortisol *Sens. Actuators* B **182** 139–46

[50] Preston A and Edgar W 2005 Developments in dental plaque pH modelling *J. Dent.* **33** 209–22

[51] Chiappin S, Antonelli G, Gatti R and Elio F 2007 Saliva specimen: a new laboratory tool for diagnostic and basic investigation *Clin. Chim. Acta* **383** 30–40

[52] Kim J, Valdés-Ramírez G, Bandodkar A J, Jia W, Martinez A G, Ramírez J, Mercier P and Wang J 2014 Non-invasive mouthguard biosensor for continuous salivary monitoring of metabolites *Analyst* **139** 1632–6

[53] Emorine M, Mielle P, Maratray J, Septier C, Thomas-Danguin T and Salles C 2012 Use of sensors to measure in-mouth salt release during food chewing *IEEE Sens. J.* **12** 3124–30

[54] Yan Q, Peng B, Su G, Cohan B E, Major T C and Meyerhoff M E 2011 Measurement of tear glucose levels with amperometric glucose biosensor/capillary tube configuration *Anal. Chem.* **83** 8341–6

[55] Nakatsukasa M, Sotozono C, Shimbo K, Ono N, Miyano H, Okano A, Hamuro J and Kinoshita S 2011 Amino acid profiles in human tear fluids analyzed by high-performance liquid chromatography and electrospray ionization tandem mass spectrometry *Am. J. Ophthalmol.* **151** 799–808. e1

[56] Choy C K M, Cho P, Chung W-Y and Benzie I F 2001 Water-soluble antioxidants in human tears: effect of the collection method *Invest. Ophthalmol. Vis. Sci.* **42** 3130–4

[57] Haeringen N V and Glasius E 1977 Collection method dependant concentrations of some metabolites in human tear fluid, with special reference to glucose in hyperglycaemic conditions *Albrecht Von Graefes Arch. Klin. Exp. Ophthalmol.* **202** 1–7

[58] Tanaka A, Utsunomiya F and Douseki T 2015 Wearable self-powered diaper-shaped urinary-incontinence sensor suppressing response-time variation with 0.3 V start-up converter *IEEE Sens. J.* **16** 3472–9

[59] Patel N D, Ward R D, Calle J, Remer E M and Monga M 2017 Vascular disease and kidney stones: abdominal aortic calcifications are associated with low urine pH and hypocitraturia *J. Endourol.* **31** 956–61

[60] Manissorn J, Fong-Ngern K, Peerapen P and Thongboonkerd V 2017 Systematic evaluation for effects of urine pH on calcium oxalate crystallization, crystal-cell adhesion and internalization into renal tubular cells *Sci. Rep.* **7** 1798

[61] Lai H-C, Chang S-N, Lin H-C, Hsu Y-L, Wei H-M, Kuo C-C, Hwang K-P and Chiang H-Y 2019 Association between urine pH and common uropathogens in children with urinary tract infections *J. Microbiol., Immunol. Infect.* **54** 290-8

[62] Maalouf N M, Cameron M A, Moe O W and Sakhaee K 2010 Metabolic basis for low urine pH in type 2 diabetes *Clin. J. Am. Soc. Nephrol.* **5** 1277–81

[63] Mitsubayashi K, Suzuki M, Tamiya E and Karube I 1994 Analysis of metabolites in sweat as a measure of physical condition *Anal. Chim. Acta* **289** 27–34

[64] Bergeron M 2003 Heat cramps: fluid and electrolyte challenges during tennis in the heat *J. Sci. Med. Sport* **6** 19–27

[65] Stern R C 1997 The diagnosis of cystic fibrosis *New Engl. J. Med.* **336** 487–91

[66] Pilardeau P, Vaysse J, Garnier M, Joublin M and Valeri L 1979 Secretion of eccrine sweat glands during exercise *Br. J. Sports Med.* **13** 118

[67] Heaney R P 1992 Calcium in the prevention and treatment of osteoporosis *J. Intern. Med.* **231** 169–80

[68] Klesges R C, Ward K D, Shelton M L, Applegate W B, Cantler E D, Palmieri G M, Harmon K and Davis J 1996 Changes in bone mineral content in male athletes: mechanisms of action and intervention effects *JAMA* **276** 226–30

[69] Gamella M, Campuzano S, Manso J, de Rivera G G, López-Colino F, Reviejo A and Pingarrón J 2014 A novel non-invasive electrochemical biosensing device for *in situ* determination of the alcohol content in blood by monitoring ethanol in sweat *Anal. Chim. Acta* **806** 1–7

[70] Marzouk S A M, Ufer S, Buck R P, Johnson T A, Dunlap L A and Cascio W E 1998 Electrodeposited iridium oxide pH electrode for measurement of extracellular myocardial acidosis during acute ischemia *Anal. Chem.* **70** 5054–61

[71] Grant S A, Bettencourt K, Krulevitch P, Hamilton J and Glass R 2001 In vitro and *in vivo* measurements of fiber optic and electrochemical sensors to monitor brain tissue pH *Sens. Actuators* B **72** 174–9

[72] Cao H, Landge V, Tata U, Seo Y S, Rao S, Tang S J, Tibbals H F, Spechler S and Chiao J C 2012 An implantable, batteryless, and wireless capsule with integrated impedance and pH sensors for gastroesophageal reflux monitoring *IEEE Trans. Biomed. Eng.* **59** 3131–9

[73] Han J, Cui D, Li Y, Zhang H, Huang Y, Zheng Z, Zhu Y and Li X 2000 A gastroesophageal tract pH sensor based on the H+-ISFET and the monitoring system for 24 h *Sens. Actuators* B **66** 203-4

[74] Nguyen C M, Huang W D, Rao S, Cao H, Tata U, Chiao M and Chiao J C 2013 Sol–gel iridium oxide-based pH sensor array on flexible polyimide substrate *IEEE Sens. J.* **13** 3857–64

[75] Nie C, Frijns A, Zevenbergen M and Toonder J D 2016 An integrated flex-microfluidic-Si chip device towards sweat sensing applications *Sens. Actuators* B **227** 427–37

[76] Schmid-Wendtner M-H and Korting H C 2006 The pH of the skin surface and its impact on the barrier function *Skin Pharmacol. Physiol.* **19** 296–302

[77] Curto V F, Fay C, Coyle S, Byrne R, O'Toole C, Barry C, Hughes S, Moyna N, Diamond D and Benito-Lopez F 2012 Real-time sweat pH monitoring based on a wearable chemical barcode micro-fluidic platform incorporating ionic liquids *Sens. Actuators* B **171–2** 1327–34

[78] Matzeu G, Florea L and Diamond D 2015 Advances in wearable chemical sensor design for monitoring biological fluids *Sens. Actuators* B **211** 403–18

[79] Mariani F, Serafini M, Gualandi I, Arcangeli D, Decataldo F, Possanzini L, Tessarolo M, Tonelli D, Fraboni B and Scavetta E 2021 Advanced wound dressing for real-time pH monitoring *ACS Sens.* **6** 2366–77

[80] Rahimi R, Ochoa M, Parupudi T, Zhao X, Yazdi I K, Dokmeci M R, Tamayol A, Khademhosseini A and Ziaie B 2016 A low-cost flexible pH sensor array for wound assessment *Sens. Actuators* B **229** 609–17

[81] Martínez-Máñez R, Soto J, García-Breijo E, Gil L, Ibáñez J and Gadea E 2005 A multisensor in thick-film technology for water quality control *Sens. Actuators* A **120** 589–95

[82] Atkinson J K, Cranny A W J, Glasspool W V and Mihell J A 1999 An investigation of the performance characteristics and operational lifetimes of multi-element thick film sensor arrays used in the determination of water quality parameters *Sens. Actuators* B **54** 215–31

[83] Koncki R and Mascini M 1997 Screen-printed ruthenium dioxide electrodes for pH measurements *Anal. Chim. Acta* **351** 143–9

[84] Liao Y-H and Chou J-C 2008 Preparation and characteristics of ruthenium dioxide for pH array sensors with real-time measurement system *Sens. Actuators* B **128** 603–12

[85] Uppuluri K, Szwagierczak D, Fernandes L, Zaraska K, Lange I, Synkiewicz-Musialska B and Manjakkal L 2023 A high-performance pH-sensitive electrode integrated with a multi-sensing probe for online water quality monitoring *J. Mater. Chem.* C **11** 15512–20

[86] Markapudi P R, Beg M, Kadara R O, Paul F, Kerrouche A, See C H and Manjakkal L 2025 Nitrate pollution mapping for reservoirs using flexible sensors integrated with underwater robot *IEEE Internet Things J.* **18** 39172–80

[87] Rabak A, Uppuluri K, Franco F F, Kumar N, Georgiev V P, Gauchotte-Lindsay C, Smith C, Hogg R A and Manjakkal L 2023 Sensor system for precision agriculture smart watering can *Res. Eng.* **19** 101297

[88] Nyein H Y Y *et al* 2016 A wearable electrochemical platform for noninvasive simultaneous monitoring of Ca^{2+} and pH *ACS Nano* **10** 7216–24

www.ingramcontent.com/pod-product-compliance
Lightning Source LLC
Chambersburg PA
CBHW080555220326
41599CB00032B/6489